Lecture Notes in Computer Scie

Commenced Publication in 1973
Founding and Former Series Editors:
Gerhard Goos, Juris Hartmanis, and Jan van Leeuwen

Editorial Board

Francesco Bonchi   Elena Ferrari
Wei Jiang   Bradley Malin (Eds.)

Privacy, Security, and Trust in KDD

Second ACM SIGKDD International Workshop, PinKDD 2008
Las Vegas, NV, USA, August 24-27, 2008
Revised Selected Papers

 Springer

Volume Editors

Francesco Bonchi
Yahoo! Research Barcelona
08018 Barcelona, Spain
E-mail: bonchi@yahoo-inc.com

Elena Ferrari
University of Insubria
Department of Computer Science and Communication
21100, Varese, Italy
E-mail: elena.ferrari@uninsubria.it

Wei Jiang
311 Computer Science Building, 500W. 15th St.
Rolla, MO 65409, USA
E-mail: wjiang@mst.edu

Bradley Malin
Vanderbilt University, Department of Biomedical Informatics
Nashville, TN 37203, USA
E-mail: b.malin@vanderbilt.edu

Library of Congress Control Number: Applied for

CR Subject Classification (1998): H.4, H.3, C.2, H.2, D.4.6, K.4-6

LNCS Sublibrary: SL 4 – Security and Cryptology

ISSN	0302-9743
ISBN-10	3-642-01717-7 Springer Berlin Heidelberg New York
ISBN-13	978-3-642-01717-9 Springer Berlin Heidelberg New York

springer.com

© Springer-Verlag Berlin Heidelberg 2009
Printed in Germany

Typesetting: Camera-ready by author, data conversion by Scientific Publishing Services, Chennai, India
Printed on acid-free paper   SPIN: 12676996   06/3180   5 4 3 2 1 0

Preface

Privacy, security, and trust in data mining are crucial and related issues that have captured the attention of many researchers, administrators, and legislators. Consequently, data mining for improved security and the study of suitable trust models, as well as data mining side-effects on privacy, have rapidly become a hot and lively research area. The issues are rooted in the real-world and concern academia, industry, government, and society in general. The issues are global, and many governments are struggling to set national and international policies on privacy, security, and trust for data mining endeavors. In industry, this is made evident by the fact that major corporations, many of which are key supporters of knowledge discovery and data mining (KDD) including IBM, Microsoft, and Yahoo!, are allocating significant resources to study and develop commercial products that address these issues. For example, at last year's PinKDD workshop, researchers from Yahoo! Research won the best paper award for their analysis of privacy issues in search queries. Beyond research, IBM has sponsored a Privacy Institute[1] and developed products, such as Hippocratic Databases[2]. These efforts have only scratched the surface of the problem, and there remain many open research issues for further investigation. For instance, the National Science Foundation recently funded the multi-institutional Team for Research in Secure Technologies[3] (TRUST) where privacy-preserving data mining is a principal focus of researchers' work in areas ranging from healthcare to wireless sensor networks. The analysis of the security, privacy, and trust aspects of data mining has begun, but they are still relatively new concepts and require workshops to promote public awareness and to present emerging research. By supporting the development of privacy-aware data mining technology, we can enable a wider social acceptance of a multitude of new services and applications based on the knowledge discovery process.

Ensuring privacy and security as well as establishing trust are essential for the provision of electronic and knowledge-based services in modern e-business, e-commerce, e-government, and e-health environments. To inject privacy and trust into security and surveillance data mining projects, it is necessary to understand what the goals of the latter are. This volume of *Lecture Notes in Computer Science* presents the proceedings of the Second International Workshop on Privacy, Security, and Trust in KDD(PinKDD 2008), which was held in conjunction with the 14th ACM SIGKDD International Conference on Knowledge Discovery and Data Mining. The workshop was held on August 24, 2008 in Las Vegas, Nevada and allowed researchers from disparate environments, including business,

---

[1] http://www.research.ibm.com/privacy/
[2] http://www.zurich.ibm.com/pri/projects/hippocratic.html
[3] http://www.truststc.org/

security, and theory to learn about the concerns and potential solutions regarding their challenges within a data mining framework.

The PinKDD 2008 workshop attracted attention from the research community and support from both industrial organizations and academic institutions. The workshop received a number of high-quality research paper submissions, each of which was reviewed by a minimum of three members of the Program and Organizing Committee. In all, six papers were presented at the workshop and five were selected for extension and inclusion in the workshop's proceedings presented in this volume. The papers represented the diversity of data mining research issues in privacy, security, and trust. In addition to two research sessions, the workshop highlights included a keynote talk which was delivered by Bhavani Thuraisingham (University of Texas at Dallas) and a panel on privacy issues in geographic data mining: the panel consisted of Peter Christen (Australian National University) and Franco Turini (University of Pisa).

December 2008   Francesco Bonchi
                                                              Elena Ferrari
                                                             Bradley Malin
                                                             Yücel Saygın

Organization

Program Chairs

Francesco Bonchi   Yahoo! Research, Spain
Elena Ferrari   University of Insubria, Italy
Wei Jiang   Missouri University of Science and
    Technology, USA
Bradley Malin   Vanderbilt University, USA

Program Committee

Maurizio Atzori   ISTI-CNR, Italy
Elisa Bertino   Purdue University, West Lafayette, USA
Barbara Carminati   University of Insubria, Varese, Italy
Peter Christen   Australian National University, Canberra,
    Australia
Christopher Clifton   Purdue University, West Lafayette, USA
Josep Domingo-Ferrer   Rovira i Virgili University, Tarragona, Spain
Tyrone Grandison   IBM Almaden Research Center, USA
Dawn Jutla   Saint Mary's University, Halifax, Canada
Murat Kantarcioglu   University of Texas, Dallas, USA
Ashwin Machanavajjhala   Cornell University, USA
Stan Matwin   University of Ottawa, Canada
Taneli Mielikäinen   Nokia Research Center, Palo Alto, USA
Yücel Saygin   Sabanci University, Istanbul, Turkey
Kian-Lee Tan   National University of Singapore
Bhavani Thuraisingham   University of Texas, Dallas, USA
Vicenç Torra   Spanish Scientific Research Council,
    Bellaterra, Spain
Vassilios Verykios   University of Thessaly, Volos, Greece
Ke Wang   Simon Fraser University, Canada
Rebecca Wright   Rutgers University, USA
Jeffrey Yu   Chinese University of Hong Kong

Table of Contents

Data Mining for Security Applications and Its Privacy Implications

Bhavani Thuraisingham

The University of Texas at Dallas
Bhavani.thuraisingham@utdallas.edu

Abstract. In this paper we first examine data mining applications in security and their implications for privacy. We then examine the notion of privacy and provide an overview of the developments especially those on privacy preserving data mining. We then provide an agenda for research on privacy and data mining.

1 Introduction

Data mining is the process of posing queries and extracting patterns, often previously unknown from large quantities of data using pattern matching or other reasoning techniques. Data mining has many applications in security including for national security as well as for cyber security. The threats to national security include attacking buildings, destroying critical infrastructures such as power grids and telecommunication systems. Data mining techniques are being investigated to find out who the suspicious people are and who is capable of carrying out terrorist activities. Cyber security is involved with protecting the computer and network systems against corruption due to Trojan horses, worms and viruses. Data mining is also being applied to provide solutions for intrusion detection and auditing.

While data mining has many applications in security, there are also serious privacy concerns. Due to data mining, even naïve users can associate data and make sensitive associations. Therefore we need to enforce the privacy of individuals while carrying out useful data mining. In this paper we will discuss developments and directions on privacy and data mining. In particular, we will provide an overview of data mining, the various types of threats and then discuss the consequences to privacy.

The organization of this paper is as follows. Section 2 discusses data mining for security applications. Overview of privacy is given in section 3. Developments on privacy aspects of data mining are discussed in section 4. Directions are given in section 5. The paper is concluded in section 6.

2 Data Mining for Security Applications

Data mining is becoming a key technology for detecting suspicious activities. In this section we discuss the use of data mining both for non real-time as well as for real-time applications. In order to carry out data mining for counterterrorism applications,

F. Bonchi et al. (Eds.): PinkDD 2008, LNCS 5456, pp. 1–6, 2009.
© Springer-Verlag Berlin Heidelberg 2009

one needs to gather data from multiple sources. For example, the following information on terrorist attacks is needed at the very least: who, what, where, when, and how; personal and business data of the potential terrorists: place of birth, ethnic origin, religion, education, work history, finances, criminal record, relatives, friends and associates, and travel history; unstructured data: newspaper articles, video clips, speeches, emails, and phone records. The data has to be integrated, warehoused and mined. One needs to develop profiles of terrorists, and activities/threats. The data has to be mined to extract patterns of potential terrorists and predict future activities and targets. Essentially one needs to find the "needle in the haystack" or more appropriately suspicious needles among possibly millions of needles. Data integrity is important and also the techniques have to SCALE.

For many applications such as emergency response, one needs to carry out real-time data mining. Data will be arriving from sensors and other devices in the form of continuous data streams including breaking news, video releases, and satellite images. Some critical data may also reside in caches. One needs to rapidly sift through the data and discard unwanted data for later use and analysis (non-real-time data mining). Data mining techniques need to meet timing constraints and may have to adhere to quality of service (QoS) tradeoffs among timeliness, precision and accuracy. The results have to be presented and visualized in real-time. In addition, alerts and triggers will also have to be implemented.

To effectively apply data mining for security applications as well as to develop appropriate tools we have to first determine what our current capabilities are. For example, do the commercial tools scale? Do they work only on special data and limited cases? Do they deliver what they promise? We need an unbiased objective study with demonstrations. At the same time, we also need to work on the big picture. For example, what do we want the data mining tools to do? What are our end results for the foreseeable future? What are the criteria for success? How do we evaluate the data mining algorithms? What test beds do we build? We need both a near-term as well as longer-term solutions. For the near-term, we need to leverage current efforts and fill the gaps in a goal-directed way as well as carry out technology transfer. For the longer-term, we need a research and development plan.

In summary, data mining is very useful to solve security problems. Tools could be used to examine audit data and flag abnormal behavior. There is a lot of recent work on applying data mining for cyber security applications, Tools are being examined to determine abnormal patterns for national security including those based on classification and link analysis. Law enforcement is also using such tools for fraud detection and crime solving.

3 Privacy Implications

Before we examine the privacy implications of data mining and propose effective solutions, we need to deter mine what is meant by privacy. In fact different communities have different notions of privacy. In the case of the medical community privacy is about a patient determining what information the doctor should release about him/her. Typically employers, marketers and insurance corporations may seek information about individuals. It is up to the individuals to determine the information to be released about him. In the financial community, a bank customer determines what financial information the bank should release about him/her. In addition, retail corporations should not be releasing the sales information about the individuals unless the individuals have authorized the release. In the case of the government community, privacy may take a whole new meaning. For example, the students who take my classes at AFCEA have mentioned to me that FBI would collect information about US citizens. However FBI determines what information about a US citizen it can release to say the CIA. That is, the FBI has to ensure the privacy of US citizens. In addition, allowing access to individual travel and spending data as well as his/her web surfing behavior should also be released upon getting permission from the individuals.

Now that we have defined what we mean by privacy, we will now examine the privacy implications of data mining. Data mining gives us "facts" that are not obvious to human analysts of the data. For example, can general trends across individuals be determined without revealing information about individuals? On the other hand, can we extract highly private associations from public data? In the former case we need to protect the individual data values while revealing the associations or aggregation while in the latter case we need to protect the associations and correlations between the data.

4   Developments in Privacy

Various types of privacy problems have been studied by researchers. We will list the various problems and the solutions proposed. (i) Problem: Privacy violations that result due to data mining: In this case the solution is Privacy-preserving data mining. That is, we carry out data mining and give out the results without revealing the data values used to carry out data mining. (ii) Problem: Privacy violations that result due to the Inference problem. Note that Inference is the process of deducing sensitive information from the legitimate responses received to user queries. The solution to this problem is Privacy Constraint Processing. (iii) Problem: Privacy violations due to unencrypted data: the solution to this problem is to utilize Encryption at different levels. (iv) Problem: Privacy violation due to poor system design. Here the solution is to develop methodology for designing privacy-enhanced systems. Below we will examine the solutions proposed for both privacy constraint/policy processing as well as for privacy preserving data mining.

Privacy constraint/policy processing research was carried out by Thuraisingham [8] and is based on some of her prior research on security constraint processing. Example of privacy constraints includes the following. Simple Constraint: an attribute of a document is private. Content-based constraint: If document contains information about X, then it is private. Association-based Constraint: Two or more documents taken together are private; individually each document is public. Release constraint: After X is released Y becomes private. The solution proposed is to augment a database system with a privacy controller for constraint processing. During the query operation the constraints are examined and only the public information is released unless of course the user is authorized to acquire the private information. Our approach also includes processing constraints during the database update as well as design operations. For details we refer to [8].

Some early work on handling the privacy problem that results from data mining was carried out by Clifton at the MITRRE Corporation [9]. The idea here is to prevent useful results from mining. One could introduce "cover stories" to give "false" results. Another approach is to only make a sample of data available so that an adversary is unable to come up with useful rules and predictive functions. However these approaches did not take off as it defeated the purpose of data mining. The goal is to carry out effective data mining but at the same time protect individual data values and sensitive associations.

Agrawal was the first to coin the term privacy preserving data mining. His initial work was to introduce random values into the data or to perturb the data so that the actual data could be protected. The challenge is to introduce random values or perturb the values without affecting the data mining results [1]. Another novel approach is the Secure Multi-party Computation (SMC) approach by Kantarcioglu and Clifton [3]. Here, each party knows its own inputs but not the others' inputs. In addition the final data mining results are also known to all. Various encryption techniques used to ensure that the individual values are protected.

SMC is showing a lot of promise and can be used also for privacy preserving distributed data mining. It is provably secure under some assumptions and the learned models are accurate; It is assumed that protocols are followed which is a semi honest model. Malicious model is also explored in some recent work by Kantarcioglu and Kardes [4]. Many SMC based privacy preserving data mining algorithms share common sub-protocols (e.g. dot product, summation, etc.). SMC does have a drawback as its not efficient enough for very large datasets. (e.g. petabyte sized datasets); Semi-honest model may not be realistic and the malicious model is even slower. There are some new directions where new models are being explored that can trade-off better between efficiency and security. Game theoretic and incentive issues are also being explored. Finally combining anonimization with cryptographic techniques is also a direction.

Before carrying out an evaluation of the data mining algorithms, one needs to determine the goals. In some cases the goal is to distort data while still preserving some properties for data mining. Another goal is to achieve a high data mining accuracy with maximum privacy protection. Our recent work assumes that Privacy is a personal choice, so should be individually adaptable. That is, we want to make privacy preserving data mining approaches to reflect reality. We investigated perturbation based approaches with real-world data sets and gave an applicability study to the current approaches [5]. We found that the reconstruction of the original distribution may not work well with real-world data sets. We tried to modify perturbation techniques as well as adapt the data mining tools. We also developed a novel privacy preserving decision tree algorithm [6].

Another development is the platform for privacy preferences (P3P) by the World Wide Web consortium (W3C). P3P is an emerging standard that enables web sites to express their privacy practices in a standard format. The format of the policies can be automatically retrieved and understood by user agents. When a user enters a web site, the privacy policies of the web site are conveyed to the user; If the privacy policies are different from user preferences, the user is notified; User can then decide how to proceed. Several major corporations are working on P3P standards.

5 Directions for Privacy

Thuraisingham proved in 1990 that the inference problem in general was unsolvable; therefore the suggestion was to explore the solvability aspects of the problem [7]. We were able to show similar results for the privacy problem. Therefore we need to examine the complicity classes as well as the storage and time complexity. We also need to explore the foundation of privacy preserving data mining algorithms and related privacy solutions. There are numerous such algorithms. How do they compare with each other? We need a test bed with realistic parameters to test the algorithms. Is it meaningful to examine privacy preserving data mining for every data mining algorithm and for every application?

It is also time to develop real world scenarios where these algorithms can be utilized. Is it feasible to develop realistic commercial products or should each organization adapt products to suit their needs? Examining privacy may make sense for healthcare and financial applications. Does privacy work for Defense and Intelligence applications? Is it even meaningful to have privacy for surveillance and geospatial applications? Once the image of my house is on Google Earth, then how much privacy can I have? I may want my location to be private, but does it make sense if a camera can capture a picture of me? If there are sensors all over the place, is it meaningful to have privacy preserving surveillance? This suggests that we need application specific privacy.

Next what is the relationship between confidentiality privacy and trust? If I as a user of Organization A send data about me to Organization B, then assume I read the privacy policies enforced by Organization B. If I agree to the privacy policies of Organization B, then I will send data about me to organization B. If I do not agree with the policies of organization B, then I can negotiate with organization B. Even if the web site states that it will not share private information with others, do I trust the web site? Note: while confidentiality is enforced by the organization, privacy is determined by the user. Therefore for confidentiality, the organization will determine whether a user can have the data. If so, then the organization can further determine whether the user can be trusted.

Another direction is how can we ensure the confidentiality of the data mining processes and results? What sort of access control policies do we enforce? How can we trust the data mining processes and results as well as verify and validate the results? How can we integrate confidentiality, privacy and trust with respect to data mining? We need to examine the research challenges and form a research agenda.

One question that Rakesh Agrawal asked at the 2003 SIGKDD panel on Privacy [2] "is privacy and data mining friends or foes? We believe that they are neither friends nor foes. We need advances in both data mining and privacy. We need to design flexible systems. For some applications one may have to focus entirely on "pure" data mining while for some others there may be a need for "privacy-preserving" data mining. We need flexible data mining techniques that can adapt to the changing environments. We believe that technologists, legal specialists, social scientists, policy makers and privacy advocates MUST work together.

6 Summary and Conclusion

In this paper we have examined data mining applications in security and their implications for privacy. We have examined the notion of privacy and then discussed the developments especially those on privacy preserving data mining. We then provided an agenda for research on privacy and data mining.

Here are our conclusions. There is no universal definition for privacy, each organization must definite what it means by privacy and develop appropriate privacy policies. Technology alone is not sufficient for privacy; we need Technologists, Policy expert, Legal experts and Social scientists to work on Privacy. Some well known people have said 'Forget about privacy" Therefore, should we pursue research on Privacy? We believe that there are interesting research problems, therefore we need to continue with this research. Furthermore, some privacy is better than nothing. Another school of thought is tried to prevent privacy violations and if violations occur then prosecute. We need to enforce appropriate policies and examine the legal aspects. We need to tackle privacy from all directions.

References

[1] Agrawal, R., Srikant, R.: Privacy-Preserving Data Mining. In: SIGMOD Conference, pp. 439–450 (2000)
[2] Agrawal, R.: Data Mining and Privacy: Friends or Foes. In: SIGKDD Panel (2003)
[3] Kantarcioglu, M., Clifton, C.: Privately Computing a Distributed k-nn Classifier. In: Boulicaut, J.-F., Esposito, F., Giannotti, F., Pedreschi, D. (eds.) PKDD 2004. LNCS, vol. 3202, pp. 279–290. Springer, Heidelberg (2004)
[4] Kantarcioglu, M., Kardes, O.: Privacy-Preserving Data Mining Applications in the Malicious Model. In: ICDM Workshops, pp. 717–722 (2007)
[5] Liu, L., Kantarcioglu, M., Thuraisingham, B.M.: The applicability of the perturbation based privacy preserving data mining for real-world data. Data Knowl. Eng. 65(1), 5–21 (2008)
[6] Liu, L., Kantarcioglu, M., Thuraisingham, B.M.: A Novel Privacy Preserving Decision Tree. In: Proceedings Hawaii International Conf. on Systems Sciences (2009)
[7] Thuraisingham, B.: One the Complexity of the Inference Problem. In: IEEE Computer Security Foundations Workshop (1990) (also available as MITRE Report, MTP-291)
[8] Thuraisingham, B.M.: Privacy constraint processing in a privacy-enhanced database management system. Data Knowl. Eng. 55(2), 159–188 (2005)
[9] Clifton, C.: Using Sample Size to Limit Exposure to Data Mining. Journal of Computer Security 8(4) (2000)

Geocode Matching and Privacy Preservation

Peter Christen

Department of Computer Science, The Australian National University
Canberra ACT 0200, Australia
peter.christen@anu.edu.au

Abstract. Geocoding is the process of matching addresses to geographic locations, such as latitudes and longitudes, or local census areas. In many applications, addresses are the key to geo-spatial data analysis and mining. Privacy and confidentiality are of paramount importance when data from, for example, cancer registries or crime databases is geocoded. Various approaches to privacy-preserving data matching, also called record linkage or entity resolution, have been developed in recent times. However, most of these approaches have not considered the specific privacy issues involved in geocode matching. This paper provides a brief introduction to privacy-preserving data and geocode matching, and using several real-world scenarios the issues involved in privacy and confidentiality for data and geocode matching are illustrated. The challenges of making privacy-preserving matching practical for real-world applications are highlighted, and potential directions for future research are discussed.

Keywords: Data matching, record linkage, entity resolution, privacy preservation, geocoding, secure multi-party computations.

1 Introduction

Increasingly large amounts of data are being collected by many business organisations, government agencies, and research institutes. A large portion of this data is about people, for example patients, students, clients and customers, travellers, or tax payers. Commonly, personal details (such as names, addresses, dates of birth, telephone, driver's license and social security numbers) are stored in databases together with application specific information, for example medical details, student enrolments and grades, customer orders and payments, travel and immigration details, or tax payments. Most of this information is considered to be private or confidential, and appropriate laws and regulations are in place in many countries that assure such information is protected properly, and that data holders do not publish such information or disclose it to others.

In our networked world, however, where many of us gather information and interact with each other and various organisations online, there is both a need and a desire for services that allow publication of information by both individuals and organisations. Examples include social networking and blogging Web sites where individuals can publish ideas, comments, profiles, photos and videos; or online mapping services that allow users to search for locations and addresses,

F. Bonchi et al. (Eds.): PinkDD 2008, LNCS 5456, pp. 7–24, 2009.

that enable the uploading of photos taken at certain locations, or that show the geographic locations of recent news events. While many of these online services are useful, informative and entertaining, they can also pose a threat to privacy, because they facilitate the way in which private or confidential pieces of information can be accessed, matched and made public.

It was recently estimated that around 80% to 90% of all governmental data collections contain details of some kind of geographic locations [1]. In many cases, these locations are addresses, and they are the key to spatially enabled data, for example to match personal information to geographic locations. The aim of such *geocode matching* is to generate geographic locations (such as latitudes and longitudes) from street address information, as will be discussed in more details in Sect. 2.1. Once data has been geocoded, it can be used for spatial data mining, and it can be visualised and combined with other data using Geographical Information Systems (GIS) and online mapping services.

In the health sector, geocoded databases can be used to detect local clusters of diseases, and many environmental studies rely upon geocoded data and GIS to map and visualise areas of possible exposure to health risks, and to locate people who live in relation to these areas. Geocoded data can help businesses to better plan marketing and future expansion, for example where to locate new stores or supply centres. In national censuses, geocoded data can be used to assign households to local census areas, which are often the basis of a variety of statistical data analysis projects.

The remainder of this paper is structured as follows. In the next section, a general overview of *data matching* is given, while *geocode matching* is discussed in Sect. 2.1. The technique of *reverse geocoding* is then described in Sect. 2.2, and the issues of privacy and confidentiality of data matching in general are the topic of Sect. 2.3. In order to illustrate these issues, several data and geocode matching scenarios are given in Sect. 3. Various privacy-preserving data matching techniques have been developed in the past few years, and an overview of these techniques is provided in Sect. 4. Finally, the paper is concluded in Sect. 5 with an outlook to potential research directions.

2   Data Matching

Data matching is the process of linking and aggregating records that refer to the same entity from one or more databases [2, 3]. A variety of techniques for data matching have been developed in different fields in the past, and while computer scientists speak of *data* or *record matching*, or *entity resolution*, statisticians and health researchers refer to *data* or *record linkage*, and the database and business oriented IT communities call this process *data cleaning* or *cleansing*, *ETL* (extraction, transformation and loading), *object identification*, or *merge/purge processing*. When the aim is to find duplicate records (i.e., records that refer to the same entity) in only one database, then this process is often called *duplicate detection*, *deduplication*, or *internal data linkage* [4].

Traditionally, data matching has been used in the health sector for matching epidemiological and administrative databases for research purposes [5], and by census agencies to create data sets that allow the production of a variety of statistical analyses [3]. Today, data matching techniques are applied in an increasing number of both private and public sector organisations. Many businesses routinely deduplicate their customer databases, and match them with data obtained from other sources, for example for marketing purposes. Government agencies match databases to detect fraud and improve outcomes in taxation, immigration and social welfare, while security agencies conduct matching of databases from a variety of sources with the objective to assemble crime and terrorism intelligence [6,7]. Another application that increasingly relies upon data matching is the assembly of bibliometric impact data, where the aim is to collect all publications of researchers and their corresponding citation counts [8].

Commonly, no unique entity identifiers are available in all the databases to be matched. Therefore, the attributes that identify entities need to be used for the matching. In many cases, these attributes are personal identifiers, such as names, addresses, dates of birth, and telephone or social security numbers. The matching process is usually challenged because real world data is dirty [9]: values commonly contain typographical errors and variations, they can be out of date or missing, and the databases to be matched might even use different coding schemes. Cleaning and standardising databases before matching is attempted is therefore a crucial pre-processing step [10], with the main tasks being the conversion of the raw input data into well defined, consistent forms, and the resolution of representation and encoding inconsistencies.

The process of matching consists of three major steps: blocking or indexing, record pair comparison, and classification of the compared record pairs.

1. **Blocking or indexing**

   When two databases are matched, potentially each record in the first database should be compared with all records in the second database. Thus, the total number of potential comparisons is of quadratic complexity. On the other hand, most of these comparisons correspond to non-matches, because the maximum number of matches can only be in the order of the number of records in the smaller of the two databases to be matched (assuming these databases do not contain duplicate records) [2]. So, while the computational efforts potentially increase quadratically with the size of the databases to be matched, the maximum number of true matches only increases linearly.

   The main performance bottleneck in the data matching process is usually the expensive detailed comparison of attribute values between records (as discussed below) [2,11], and therefore it is not feasible to compare all record pairs when the databases are large.

   Techniques commonly know as *blocking* [11] are therefore applied, with the aim to reduce the actual number of record pairs that are to be compared. These techniques work by indexing or clustering the databases into blocks, such that candidate pairs of records, which are likely to correspond to a true match, are inserted into the same block. Pairs of records are then generated only from the records that are in the same block. While blocking will remove many of the obvious non-matching record pairs, some true matches will likely also be removed in the blocking step, because of errors or variations in record attribute values [2].

2. **Record pair comparisons**
   Selected attributes (or fields) of the candidate record pairs generated in the blocking step are compared using a variety of comparison functions [12, 13]. To account for typographical errors and variations, strings are normally compared using approximate string comparison functions [14], while specific comparison functions exist for other data types, such as ages, times, dates, or telephone, social security or credit card numbers [12].

   All these similarity functions return a numerical *matching weight*, which is commonly normalised, such that an exact match between two attribute values returns 1.0, while the comparison between two completely different values returns 0.0. For each compared record pair, a *weight vector* is generated, which contains one matching weight for each compared attribute. For example, if surnames, given names and postcodes are compared, then three matching weights would be calculated for each compared record pair.

3. **Record pair classification**
   In the traditional data matching approach [3], for each compared record pair, the weights in its corresponding weight vector are summed into one total matching weight. Using two thresholds, a record pair is then classified as a *match*, a *non-match*, or as a *possible match* (these are the record pairs for which manual *clerical review*, a time-consuming and cumbersome process, is required to decide their final match status). In the past decade, researchers from the fields of data mining, machine learning, artificial intelligence and databases have explored the use of various techniques with the objective to make the classification of record pairs both more accurate and more automatic [2,3]. Many of these novel techniques are based on supervised machine learning approaches and therefore require training data, which unfortunately is often not available in real world applications [15].

Matching today's increasingly large databases has several challenges. First, as already mentioned, real-world data is dirty, and matching therefore is reliant upon the attributes available in common in all databases to be matched. In real-world applications, data quality can be one of the biggest obstacles to successful data matching [16]. Second, even when efficient blocking techniques are applied, the number of candidate record pairs can be very large, and the resulting computational requirements (run-times and memory foot-prints) can become challenging even for today's powerful computing platforms.

Finally, because the matching is often based on personal information, like name and address details, privacy and confidentiality become of paramount concern, especially when databases are matched between different organisations. This challenge will be discussed in more detail in Sect. 2.3.

2.1   Geocode Matching

Geocode matching, or *geocoding*, is the process of matching user address data with a reference database that contains cleaned and standardised addresses and their geographic locations [17]. Geocoding is a special case of data matching, and it is challenged by the same issues as data matching, as discussed above. Specifically, addresses are often dirty, in that they can include misspellings of street and suburb names, can have wrong or incomplete street or apartment numbers, wrong zipcodes, or parts of an address can even be missing. It has been reported that a geocode matching rate of around 70% is acceptable when user addresses have been recorded over the telephone, have been scanned, or were manually typed from hand-written forms [18].

Additionally, addresses can quickly become out-of-date as people move to new homes. This is especially an issue with younger people who are much more mobile than older generations. As new suburbs are being built, and postal agencies adjust their zipcode areas, new addresses are becoming valid, while parts of existing addresses are changing and some addresses even become invalid. Matching a user address to the correct reference address can therefore become quite challenging. Scalability and computational efforts are also of concern, as a reference address database usually contains many million addresses, and so often does a user's database. Efficient blocking techniques are thus important.

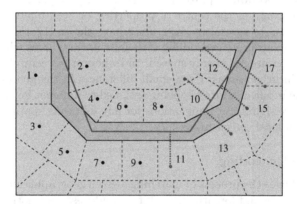

Fig. 1. Example geocoding using property centres (numbers 1 to 9) and street centrelines (the thick dark/red lines and numbers 10 to 17, with the dotted lines corresponding to a global street offset)

As illustrated in Fig. 1, there are two basic geocoding techniques, which depend upon the available reference database. Street level geocoding is based on a reference database that contains street centreline data, which are the geographic locations of the start- and end-points of street segments, as well as the range of street numbers they cover. A popular, freely available such database for the USA is the *Tiger* (Topologically Integrated Geographic Encoding and Referencing)

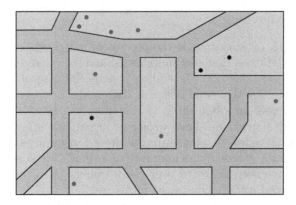

Fig. 2. Reverse-geocoding example. The dots on the map could, for example, correspond to certain disease cases or the locations of certain crime types.

system by the US Census Bureau.[1] With this type of geocoding, positional errors are introduced because street segments normally do not follow the exact location of streets, the locations of street numbers are calculated assuming regular intervals between properties, and commonly a global street offset (distance from the street centreline) is used to calculate the actual geographic locations.

An alternative geocoding approach is possible if a property reference database is available that contains the centre of properties, or even buildings, for all addresses in an area. As several recent studies have shown [17, 19], this approach can result in more accurate geocoding, especially in rural areas where properties and street numbers are not evenly spaced, and where street centreline data is less accurate than in urban areas. Even small differences in the geocoded locations can for example result in addresses being assigned to different local census areas, and this can have significant implications for any following spatial analysis that is based on such local census areas [17].

2.2   Reverse Geocoding

Reverse geocoding is the process of matching a given geographic location (usually provided as latitude and longitude) back to an address (or another entity, such as a building or property). The geographic location can be extracted from maps, such as the example shown in Fig. 2, which can often be taken from Web sites or from publications. One challenge with reverse geocoding lies in accurately positioning the map relative to its true geographic location to enable accurate matching of locations to addresses. Another issue is related to the resolution of a given map, i.e., the physical area covered by a single pixel, and the way locations are marked on a map (for example as single highlighted pixels, as coloured circles, or as much larger illustrative markers).

Two recent studies [20, 21] illustrated the accuracy of reverse geocoding. They were based on maps with a resolution commonly used in publications, such as

---

[1] Available from: http://www.census.gov/geo/www/tiger/

medical journals. With a higher resolution image file (226 dots per inch, and a scale of 1:100,000), the authors were able to accurately identify 432 of 550 (79%) randomly selected addresses in Boston within 14 meters of the real address [21]. With a lower resolution image file (50 dots per inch and scale 1:100,000), 144 of 550 (26%) addresses could be identified directly, while on average the real addresses were only 21 meters away from the reverse geocoded addresses. Overall, 99.8% of all addresses were within 70 meters of their real locations [20].

A similar analysis was conducted after hurricane Katrina using a map published by a local newspaper [22]. This map showed the locations of mortalities, and applying reverse geocoding, the authors were able to identify the actual locations of many of these mortalities. Using field teams with GPS receivers that recorded actual locations allowed the authors to measure the accuracy of their reverse geocoding process. They found that up to 40% of the reverse geocoded mortalities were located within 20 meters of the field verified actual residencies.

2.3   Privacy and Confidentiality Issues

Data and geocode matching are technologies that can be employed in a variety of applications. For example, the aim of matching health databases is normally to investigate the effects of various aspects of the health system and how they interact with environmental or social factors, with the objectives to increase our understanding of these complex interactions, to improve the public health system, and to reduce costs to both patients and governments. Thus, this kind of data matching is aimed at increasing the general public's wellbeing. Similarly, when data matching is used by statistical agencies as part of the census, or by other government agencies to detect fraudulent behaviour, such as people registering for unemployment money while they are actually employed, then the public is generally supportive of these types of data matching projects.

On the other hand, when data matching is used by national security agencies, for example to assemble terrorism watch lists, then the public increasingly worries about the resulting privacy and confidentiality implications, because individuals can directly be affected [6]. A false match can result in an innocent individual being added to a list of suspected terrorists or criminals, resulting in serious implications.

To a similar degree, the public is also worried about how private organisations are selling, exchanging and matching their customer data, in order to improve their business or target specific groups for marketing. In many countries, consumers have only limited control over what private organisations can do with the data they have collected about them during their business activities.

Data matching techniques can also be used by criminals, for example for matching disparate pieces of information they collect as part of their efforts to conduct identity fraud. With an increasing amount of data about various aspects of people's life being available online, matching such disparate data with the aim to build somebody's identity becomes increasingly attractive, as this can be done anywhere in the world with relatively little efforts, and does not require physical access to a victim's information (like searching through somebody's garbage with the aim to find identifying details) [23]. Such *re-identification* of individuals from using only publicly available, de-identified (or partially identifying) data has recently attracted interests from several research fields, and a variety of anonymisation techniques for data publishing have been proposed [24, 25, 26].

Geocode matching, and especially reverse geocoding, can have serious privacy implications as well. As the studies described in Sect. 2.2 showed, it can allow matching of information about people to the location where they live, potentially indicating an individual's vulnerability, such as somebody living alone or having a certain illness. Even when areas and not individual addresses are associated with certain characteristics, for example higher prevalence of a certain disease or higher crime rates, then this will affect all people living in such an area, and can result in residents and businesses avoiding or leaving the area. Health research and crime analysis are two domains that increasingly publish maps and images with geocoded data in research publications or on Web sites.

It is therefore paramount that the personal details of individuals are protected when data is matched or geocoded [27, 28]. This is especially important in applications where a breach of privacy can seriously impact an individual's financial, employment, insurance and social status. Data matching and geocoding will only receive widespread acceptance if privacy and confidentiality of the individuals whose data is being matched can be guaranteed. A variety of such approaches have recently been developed, as will be discussed in Sect. 4.

3   Data and Geocode Matching Scenarios

In this section, the previously discussed privacy and confidentiality issues are illustrated using several scenarios. The first two have been adapted from [27, 29], while the third scenario is based on ideas given in [30].

Scenario 1: Data Matching
A public health researcher is planning to investigate the effects that serious car accidents have upon the public health system. She is interested in what the most common types of injuries are, how much they cost the public health system, and how the general health of car accident victims is progressing after their recovery. For her analysis, the researcher requires access to data from public and private hospitals, general practitioners and specialist doctors, private health insurance companies, the police, and car insurance companies.

In most countries it is very unlikely that all these databases have a common entity identifier that enables exact matching of all records that refer to the same individual. Therefore, for this project, personal details are required for the matching. If the researcher is successful in getting access to all the required databases, then the matching can be performed by the researcher (or a support entity at the researcher's organisation) following strict security and access procedures. Alternatively, all databases could be given to a trusted organisation, such as a data matching unit within a government health department, which would perform the matching and only provide the matched records without any identifying information (name and address details) to the researcher.

In both these cases, however, the original records (including the identifying personal details) from all databases required for this project have to be made available to the organisation that performs the matching. This requirement will very likely prevent some of the organisations involved from providing their data towards this project, and therefore prevent a research project that is of significant benefit to the public.

Scenario 2: Geocode Matching [2]

In many countries, cancer registries collect data about all occurrences of cancer in a certain state or territory. Often, these registries are small organisations that are partially supported by government funding and public donations.

In this scenario, it is assumed a cancer registry would like to geocode its database to conduct spatial analyses of its cancer data. This would, for example, allow the registry to detect if there are clusters of cancers in certain areas, or correlations between increased cancer occurrences and environmental or socio-economic factors. Due to its limited financial resources, the cancer registry cannot afford to invest in an in-house geocoding system (consisting of software, as well as trained personnel), but instead is required to employ an external geocoding service provider.

The regulatory or legal framework under which the cancer registry operates might not allow it to provide its detailed data to any external organisation for geocoding. Even if this would be allowed, complete trust is required in the capabilities of the external geocoding service provider to accurately match the registry's data, and to then properly destroy all data received by the cancer registry or generated during the geocoding project. Additionally, if the geocoding service provider is a commercial organisation, then normally only limited independent information is available about its matching quality. Ideally, comparative studies, conducted by an independent organisation like a government agency, about the matching quality of various commercial geocoding products and service providers should be available to help the cancer registry select a suitable geocoding provider.

In order to obfuscate their data, the cancer registry might use *chaffing* [32], by adding dummy address records into its cancer database before sending it to the geocoding service provider. This will however increase the costs of geocoding, as a commercial provider will likely charge according to the number of addresses to be geocoded. As an alternative, the cancer registry might use the geocoding service of a trusted proxy organisation, such as a government health department. However, in both approaches, the original address details of the cancer patients have to be made available to the outside organisation that performs the geocoding, and this might pose a serious privacy risk.

Scenario 3: Reverse Geocoding

A police department is publishing crime statistics of its local area on its newly designed Web site. This Web site allows the public to select a variety of crime

---

[2] See [31] for a more general discussion of geocoding in cancer research.

types (like burglaries, traffic offences, assaults, etc.) and time periods when they happened. It also allows some limited drill-down into smaller areas, such as suburbs. These crime maps initially improve the relationship between the police and the public, because they raise awareness where problematic neighbourhoods are, and because they allow the public to know where crimes and security events occur. As a result, the public gains confidence in the work the police is doing, and people are also more willing to work with the police.

On the other hand, these published crime maps are also misused. For example, vendors of burglary alarms target local areas with higher incident rates of burglaries, while the property values in these areas have dropped significantly since the crime maps have been published, because it is harder to attract new residents into these areas. Some residents even leave these areas because of the published crime statistics, and as a result, local businesses also consider to move to different, safer locations.

The published crimes maps also allow criminals to identify areas that have not received a lot of police attention in recent times, for example because the crime incident rates in these areas are low. Once criminals obtain socio-economic data for these areas from a government census agency, they find that some of these areas are suburbs with high average income and a large number of single residents. This indicates to these criminals that there is an increased number of potentially easy and lucrative targets in these areas.

Besides summarised crime statistics, the police Web site also allows the generation of 'pin' maps (as illustrated in Fig. 2), which show the exact locations of crimes in recent times. Filtering and selection options allow visualisation of serious crimes, such as assaults, murders and sexual offences. This results in several victims being contacted by media organisations, which re-traumatises them. Following from these incidents, the police department sees a significant reduction in the number of reports in sexual offences, as victims decide not to report an assault because of fear of publicity.

4   Privacy-Preserving Matching

Traditionally, when databases from different organisations are to be matched, then the organisation that undertakes the matching will require access to all the records from all the databases to be matched, because it is impossible to know before the matching which records will match. Complete trust is required in the intentions of all organisations involved in a matching project, in their ability to maintain confidentiality, in the security of their networking and computing facilities, and in the reliability of the staff involved in matching projects.

In situations were the data is matched for a research project, for example, the matching is often conducted by a trusted organisation, such as a government regulated matching unit. For the matching of health related databases, good practise dictates that all medical details are removed from the data before it is given to such a matching unit [5, 33], and that the researcher is only given the required medical details of the matched records, but not the full identifying

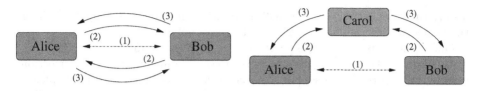

Fig. 3. Basic two- and three-party protocols for privacy-preserving data matching. Alice and Bob are the data holders, and Carol is the trusted data matching unit.

attributes. With this approach, however, even only knowing the source of the data to be matched might permit highly confidential information to be inferred about individuals who are identified in the databases to be matched.

However, this invasion of privacy could be reduced or even avoided if there were some approach that allows the detection of records that refer to the same entity in databases from different organisations, without either organisation having to reveal any identifying details to any other organisation. Such approaches have been developed in the past few years, most of them based on cryptographic and secure multi-party computation techniques.

Pioneering work was done in the 1990s by French health researchers [34,35,36, 37,38]. Their approach was based on keyed one-way hash encoding functions [39], which allowed matching using only encoded identifying data. These approaches provided good privacy protection against a single organisation trying to find out information about the other organisation's data. However, a major problem when using hash-encoded values (for example, the original un-encoded value *'tim'* could be encoded into the hash-string *'51d3a6a70'*) is that a single character difference in an original value results in a completely different hash encoding, and thus only exact matches will be found. Applying phonetic encoding, such as Soundex or NYSIIS [14], to the original values before they are hash-encoded can help to overcome some variations, but only to a limited degree.

Cryptography based privacy-preserving data matching approaches can generally be classified into two- and three-party protocols, as illustrated in Fig. 3. Within a two-party protocol, the two data holders, *Alice* and *Bob*, plan to match their databases such that only information about the matched records is revealed to each other. The following principle three steps are involved: (1) The data holders agree upon a secret random key which will be used to encode their data in the following steps. It is also assumed that all communication is conducted in an authenticated and secure fashion, for example using a public key infrastructure (PKI) [39]. (2) The two data holders pre-process and encode their databases in an agreed fashion, and then send their encoded database to each other. (3) Each data holder performs the matching using their own data and the encoded data they received from the other data holder, and thus both data holders learn which records match. The two data holders can then exchange their matching results and negotiate how to proceed next. Steps (2) and (3) might be repeated several times, depending upon the actual matching technique employed. The most important requirement for any two-party data matching protocol is that at any time neither data holder will have all the information required to infer the values of the original data of the other data holder.

One recently proposed two-party protocol for string distances is based on a stochastic scalar product, that is as secure as the underlying set-intersection cryptographic protocol it uses [40]. Another approach is aimed at secure sequence comparisons [41], and is based on edit-distance calculations. It applies encoding such that neither party at any time has the complete information about the dynamic-programming matrix used for the edit-distance calculation.

Three-party protocols for privacy-preserving data matching are based on the idea that a trusted third party, commonly called *Carol*, conducts the matching in such a way that neither of the two data holders have to provide any private or confidential information to any other party involved in the protocol, and *Carol* only sees encoded values. The general three-party protocol also consists of three principal steps, as illustrated in Fig. 3: (1) The two data holders again agree upon a secret random key which will be used to encode their data in the following steps. Note that this key is not shared with *Carol*. (2) The two data holders now pre-process and encode their databases in an agreed fashion, and then send their encoded database to *Carol*, which performs the matching without seeing any of the original values. That is, *Carol* must perform the matching on the encoded data she received from *Alice* and *Bob*. (3) Once the matching is completed, *Carol* sends information about the matched records back to both data holders. Depending upon what was agreed, this might only be the number of matched records, or it might be their record identifiers. Depending upon the outcome of the matching, *Alice* and *Bob* then negotiate how to proceed next.

Several three-party protocols have recently been developed. They mainly differ in the way the matching party is calculating the similarity between encoded record attribute values, and by how much information can be inferred by any of the parties involved in the matching protocol.

Two protocols that not only enable data matching, but also allow cohort extraction, have recently been proposed [42]. Combined, they facilitate the construction of a matched data set in such a way that no identifying information is revealed to any other party involved, and neither of the data holders learns which of their records have been extracted from their databases. These two protocols are based on hash-encoded values. They can only perform exact matching, and are thus of limited use when the data contains errors and variations.

Another three-party protocol, named *blindfolded record linkage* [43], is based on hash-encoded q-grams (sub-strings of length q), and allows approximate matching by calculating the *Dice* co-efficient similarity measure between hash-encoded sets of q-grams. The major drawback of this approach is its computational and communication overhead, which makes it impractical for large databases or long strings, such as suburb names or genome sequences.

As discussed in Sect. 2, one major challenge when matching large databases is the potential number of record pairs that need to be compared. Blocking techniques are required to make large-scale matching possible [11], however, none of the so far presented privacy-preserving matching approaches take blocking into account. A recently presented approach to secure blocking [44] is based on a three-party protocol, and uses hash-encoded values and secure string distance calculations, similar as used in other approaches [40]. The basic idea of the approach is to compare records between databases only if they have at least one token in common (for example a word or q-gram).

A different approach that was recently proposed [45] is aimed at improving the performance of privacy-preserving data matching by using a hybrid technique which combines sanitisation techniques (like k-anonymity [26]) with cryptographic secure multi-party computations (SMC). The idea of this approach is to first use the anonymised data sets to decide the match status (match or non-match) of a large proportion of record pairs, and secondly to only use SMC for the remaining, hard to match pairs. The first step of this approach can be seen as a blocking step.

4.1 Privacy-Preserving Geocoding

To the best of the author's knowledge, only one publication has so far considered privacy-preserving geocode matching [43]. What is required is that, besides the data holder, no party involved in a privacy-preserving geocode matching protocol should be able to learn which addresses were matched, as otherwise the geographic locations of the data holder's addresses would be revealed. Thus, a method for privacy-preserving geocode matching should allow a data holder to locally encode their address records and then transfer them to a geocoding service provider, without having to reveal any of their addresses, and without the geocoding service provider learning anything about these addresses.

This process is similar to the cohort extraction protocol discussed above [42], however, because addresses are often dirty (i.e., contain errors and variations), approximate matching techniques are required. A variation of the q-gram based three-party protocol [43] does allow such a privacy-preserving geocoding, but as discussed before, the computational and communication overheads of this approach prohibits the geocoding of large databases.

Besides variations and errors in user address data, geocoding, when based on a property reference address database, also has to deal with issues such as user street addresses that are not available in the reference database. In such cases, the location of the missing street number needs to be extrapolated using the closest street numbers that exist in the reference database. Similarly, if a given user address cannot be found in its expected zipcode or suburb area, the matching should be extended to neighbouring areas, because people commonly provide neighbouring zipcodes, especially if they have a higher social status [17]. Many of the approaches to privacy-preserving data matching presented in Sect. 4 can be used as starting points to develop privacy-preserving geocode matching.

4.2 Challenges

Many of the presented approaches to privacy-preserving data matching are in a proof-of-concept or prototype state, and they currently only allow matching of small to medium sized databases, or they can only perform exact matching. Before privacy-preserving data and geocode matching can be employed in practice, the following challenges need to be addressed.

– **Improved secure matching techniques**
  Approaches that enable approximate matching of databases in a privacy-preserving way have only been developed in the past few years [40,41,43,44]. Using secure multi-party computations, they securely compute functions at the expense of computational and communication overheads. Many of these approaches are of quadratic or even higher complexity, and are therefore of limited scalability. Additionally, while they allow approximate matching, they are only partial solutions, in that they are not integrated into either the traditional probabilistic data matching approach, nor into one of the various recently developed machine learning based techniques [3, 29].

– **Automated record pair classification and quality assessment**
  A variety of novel data matching classification methods has been developed in the past few years [2, 3, 15], however, none of them considers privacy preservation. Many of these advanced classification methods are based on supervised machine learning approaches, and this will make it difficult to integrate them into a privacy-preserving framework, because the party undertaking the matching in such a case usually does not have access to the original, un-encoded data values that are used for the matching. Any classifier that requires (manually prepared) training data thus becomes cumbersome, if not impossible. Unsupervised classification techniques [15, 46] are thus required that do not rely upon the original, un-encoded data values.

  Related to the need for automated classification techniques for record pairs is the challenge of how to assess the quality of the resulting matched records within a privacy-preserving framework. If the two data holders do not reveal any of their original, un-encoded record values to each other, how can one be assured that no truly matched record pairs were missed (i.e., how can recall be measured)? And of the matched records, given the party undertaking the matching does not see the original, un-encoded record values, how can precision be measured (i.e., how many of the classified matches correspond to true matches)? How to assess the quality of the matched data [2] within a privacy-preserving framework needs to be investigated when automated record pair classification techniques are being developed.

– **Scalability to large databases and real-time matching**
  Besides improved secure matching and automated record pair classification techniques, scalability to very large databases is one of the major challenges of current approaches to privacy-preserving data matching. The computational and communication overheads of most approaches developed so far prohibit the matching of large databases that contain many millions of records. What is required are novel techniques that scale linearly with the number of records to be matched, and also techniques that take advantage

of the parallel multi-core computing capabilities that increasingly become available on many modern computing platforms [47].

A related challenge is the capability to match very large databases with a stream of incoming query records in (near) real-time [48], because real-time matching in a privacy-preserving framework is becoming increasingly important. Examples include identity verification for credit card applications and matching of crime and terrorism databases for national security. Real-time geocoding is vital when data from a possible bio-terrorism attack needs to be geocoded to obtain the locations where victims are living and working, as in such situations it is crucial to know who has been in contact with a victim.

– **Preventing re-identification**
While privacy-preserving data and geocode matching assures that no data is being revealed to any of the parties involved in a matching project, or to any external attacker, the question still remains how the information about the matched records is being further handled. For example, even if only zipcode, gender and age values are being released to the researcher who is conducting a study using the matched data, then it will likely be possible for this researcher to use these three attributes and match them with other, external data that is publicly available, and then re-identify the people in the matched data set [26]. This obviously can lead to a loss of privacy and confidentiality for the individuals whose records have been re-identified.

Traditionally, work on methods that prevent re-identification has been done by statisticians [49], while recently computer scientists have also started to investigate this challenge [26, 50]. Privacy-preserving matching will only become practically relevant if it is combined with anonymisation techniques that can guarantee that no re-identification is possible in any case, even when the matched records would be further matched with additional data.

5   Conclusions and Research Directions

In this paper, an overview of data and geocode matching has been presented, the principal steps involved in data matching have been discussed, and the specific techniques used in geocode matching and reverse geocoding have been highlighted. Using several scenarios, the privacy and confidentiality issues that arise when data from different organisations is being matched or geocoded have been illustrated. An overview of the recently developed privacy-preserving matching and geocoding approaches has been provided, and the major challenges in privacy-preserving data and geocode matching have been discussed.

Future research direction should address the challenges described in Sect. 4.2, with the objective to make privacy-preserving data and geocode matching more practical. While partial solutions exist to all of the described challenges, to the best of the author's knowledge no currently available privacy-preserving data matching approach is tackling all of them.

Specifically, the computational and communication overheads of current approaches should be reduced to allow matching of very large databases, and the currently available privacy-preserving matching approaches should be integrated with advanced machine learning based classification methods to enable automated and accurate matching. Finally, techniques that are specific to geocode matching should be developed, to allow privacy-preserving geocoding in situations where a data holder cannot geocode its data otherwise.

References

1. US Federal Geographic Data Committee. Homeland Security and Geographic Information Systems: How GIS and mapping technology can save lives and protect property in post-September 11th America. Public Health GIS News and Information (52), 21–23 (May 2003)
2. Christen, P., Goiser, K.: Quality and complexity measures for data linkage and deduplication. In: Guillet, F., Hamilton, H.J. (eds.) Quality Measures in Data Mining. Studies in Computational Intelligence, vol. 43, pp. 127–151. Springer, Heidelberg (2007)
3. Winkler, W.E.: Overview of record linkage and current research directions. Technical Report RRS2006/02, US Bureau of the Census (2006)
4. Elmagarmid, A.K., Ipeirotis, P.G., Verykios, V.S.: Duplicate record detection: A survey. IEEE Transactions on Knowledge and Data Engineering 19(1), 1–16 (2007)
5. Kelman, C.W., Bass, J.A., Holman, D.: Research use of linked health data – A best practice protocol. ANZ Journal of Public Health 26(3), 251–255 (2002)
6. Jonas, J., Harper, J.: Effective counterterrorism and the limited role of predictive data mining. Policy Analysis (584) (2006)
7. Wang, G., Chen, H., Xu, J.J., Atabakhsh, H.: Automatically detecting criminal identity deception: An adaptive detection algorithm. IEEE Transactions on Systems, Man and Cybernetics (Part A) 36(5), 988–999 (2006)
8. Bhattacharya, I., Getoor, L.: Collective entity resolution in relational data. ACM Transactions on Knowledge Discovery from Data (TKDD) 1(1) (2007)
9. Hernandez, M.A., Stolfo, S.J.: Real-world data is dirty: Data cleansing and the merge/purge problem. Data Mining and Knowledge Discovery 2(1), 9–37 (1998)
10. Churches, T., Christen, P., Lim, K., Zhu, J.: Preparation of name and address data for record linkage using hidden Markov models. BioMed Central Medical Informatics and Decision Making 2(9) (2002)
11. Baxter, R., Christen, P., Churches, T.: A comparison of fast blocking methods for record linkage. In: ACM KDD Workshop on Data Cleaning, Record Linkage and Object Consolidation, Washington, DC (2003)
12. Christen, P.: Febrl – An open source data cleaning, deduplication and record linkage system with a graphical user interface. In: ACM International Conference on Knowledge Discovery and Data Mining, Las Vegas, pp. 1065–1068 (2008)
13. Cohen, W.W., Ravikumar, P., Fienberg, S.E.: A comparison of string distance metrics for name-matching tasks. In: IJCAI Workshop on Information Integration on the Web, Acapulco, pp. 73–78 (2003)
14. Christen, P.: A comparison of personal name matching: Techniques and practical issues. In: IEEE ICDM Workshop on Mining Complex Data, Hong Kong, pp. 290–294 (2006)
15. Christen, P.: Automatic record linkage using seeded nearest neighbour and support vector machine classification. In: ACM International Conference on Knowledge Discovery and Data Mining, Las Vegas, pp. 151–159 (2008)

16. Clarke, D.: Practical introduction to record linkage for injury research. Injury Prevention 10, 186–191 (2004)
17. Christen, P., Willmore, A., Churches, T.: A probabilistic geocoding system utilising a parcel based address file. In: Williams, G.J., Simoff, S.J. (eds.) Data Mining. LNCS (LNAI), vol. 3755, pp. 130–145. Springer, Heidelberg (2006)
18. Paull, D.: A geocoded national address file for Australia: The G-NAF what, why, who and when? PSMA Australia Limited, Griffith, ACT, Australia (2003), http://www.g-naf.com.au/
19. Cayo, M.R., Talbot, T.O.: Positional error in automated geocoding of residential addresses. International Journal of Health Geographics 2(10) (2003)
20. Brownstein, J.S., Cassa, C., Kohane, I.S., Mandl, K.D.: Reverse geocoding: Concerns about patient confidentiality in the display of geospatial health data. In: AMIA Annual Symposium Proceedings 2005, p. 905 (2005)
21. Brownstein, J.S., Cassa, C., Mandl, K.D.: No place to hide–reverse identification of patients from published maps. New England Journal of Medicine 355(16), 1741–1742 (2006)
22. Curtis, A.J., Mills, J.W., Leitner, M.: Spatial confidentiality and GIS: Re-engineering mortality locations from published maps about Hurricane Katrina. International Journal of Health Geographics 5(1), 44–56 (2006)
23. Australian Attorney-General's Department, Standing Committee of Attorney's-General: Model criminal law officers' committee: Final report on identity crime. Canberra (March 2008)
24. Chaytor, R., Brown, E., Wareham, T.: Privacy advisors for personal information management. In: SIGIR Workshop on Personal Information Management, Seattle, Washington, pp. 28–31 (2006)
25. Fienberg, S.E.: Privacy and confidentiality in an e-Commerce world: Data mining, data warehousing, matching and disclosure limitation. Statistical Science 21(2), 143–154 (2006)
26. Sweeney, L.: K-anonymity: A model for protecting privacy. International Journal of Uncertainty, Fuzziness and Knowledge-Based Systems 10(5), 557–570 (2002)
27. Christen, P.: Privacy-preserving data linkage and geocoding: Current approaches and research directions. In: IEEE ICDM Workshop on Privacy Aspects of Data Mining, Hong Kong, pp. 497–501 (2006)
28. Sweeney, L.: Privacy-enhanced linking. ACM SIGKDD Explorations 7(2), 72–75 (2005)
29. Christen, P., Churches, T.: Secure health data linkage and geocoding: Current approaches and research directions. In: National e-Health Privacy and Security Symposium, Brisbane, Australia (2006)
30. Wartell, J., McEwen, T.: Privacy in the information age: A guide for sharing crime maps and spatial data. Institute for Law and Justice, NCJ 188739 (July 2001)
31. Rushton, G., Armstrong, M.P., Gittler, J., Greene, B.R., Pavlik, C.E., West, M.M., Zimmerman, D.L.: Geocoding in cancer research – A review. American Journal of Preventive Medicine 30(2S), 16–24 (2006)
32. Rivest, R.L.: Chaffing and winnowing: Confidentiality without encryption. MIT Lab for Computer Science (1998), http://theory.lcs.mit.edu/~rivest/chaffing.txt
33. Churches, T.: A proposed architecture and method of operation for improving the protection of privacy and confidentiality in disease registers. BioMed. Central Medical Research Methodology 3(1) (2003)
34. Bouzelat, H., Quantin, C., Dusserre, L.: Extraction and anonymity protocol of medical file. In: AMIA Fall Symposium, pp. 323–327 (1996)

35. Dusserre, L., Quantin, C., Bouzelat, H.: A one way public key cryptosystem for the linkage of nominal files in epidemiological studies. Medinfo. 8(644–7) (1995)
36. Quantin, C., Bouzelat, H., Allaert, F.A., Benhamiche, A.M., Faivre, J., Dusserre, L.: Automatic record hash coding and linkage for epidemiological follow-up data confidentiality. Methods of Information in Medicine 37(3), 271–277 (1998)
37. Quantin, C., Bouzelat, H., Allaert, F.A., Benhamiche, A.M., Faivre, J., Dusserre, L.: How to ensure data quality of an epidemiological follow-up: Quality assessment of an anonymous record linkage procedure. International Journal of Medical Informatics 49(1), 117–122 (1998)
38. Quantin, C., Bouzelat, H., Dusserre, L.: Irreversible encryption method by generation of polynomials. Medical Informatics and the Internet in Medicine 21(2), 113–121 (1996)
39. Schneier, B.: Applied cryptography: Protocols, algorithms, and source code in C, 2nd edn. John Wiley & Sons, Inc., New York (1995)
40. Ravikumar, P., Cohen, W.W., Fienberg, S.E.: A secure protocol for computing string distance metrics. In: IEEE ICDM Workshop on Privacy and Security Aspects of Data Mining, Brighton, UK (2004)
41. Atallah, M.J., Kerschbaum, F., Du, W.: Secure and private sequence comparisons. In: ACM Workshop on Privacy in the Electronic Society, Washington DC, pp. 39–44 (2003)
42. O'Keefe, C.M., Yung, M., Gu, L., Baxter, R.: Privacy-preserving data linkage protocols. In: ACM Workshop on Privacy in the Electronic Society, Washington DC, pp. 94–102 (2004)
43. Churches, T., Christen, P.: Some methods for blindfolded record linkage. BioMed. Central Medical Informatics and Decision Making 4(9) (2004)
44. Al-Lawati, A., Lee, D., McDaniel, P.: Blocking-aware private record linkage. In: International Workshop on Information Quality in Information Systems, Baltimore, pp. 59–68 (2005)
45. Inan, A., Kantarcioglu, M., Bertino, E., Scannapieco, M.: A hybrid approach to private record linkage. In: IEEE International Conference Data Engineering, pp. 496–505 (2008)
46. Christen, P.: Automatic training example selection for scalable unsupervised record linkage. In: Washio, T., Suzuki, E., Ting, K.M., Inokuchi, A. (eds.) PAKDD 2008. LNCS (LNAI), vol. 5012, pp. 511–518. Springer, Heidelberg (2008)
47. Guisado-Gamez, J., Prat-Perez, A., Nin, J., Muntes-Mulero, V., Larriba-Pey, J.L.: Parallelizing record linkage for disclosure risk assessment. In: Privacy in Statistical Databases, Istanbul, Turkey. LNCS, vol. 5262, pp. 190–202. Springer, Heidelberg (2008)
48. Christen, P., Gayler, R.: Towards scalable real-time entity resolution using a similarity-aware inverted index approach. In: AusDM 2008, CRPIT, Glenelg, Australia, vol. 87, pp. 51–60 (2008)
49. Winkler, W.E.: Masking and re-identification methods for public-use microdata: Overview and research problems. In: Domingo-Ferrer, J., Torra, V. (eds.) PSD 2004. LNCS, vol. 3050, pp. 216–230. Springer, Heidelberg (2004)
50. Malin, B., Sweeney, L.: A secure protocol to distribute unlinkable health data. In: American Medical Informatics Association 2005 Annual Symposium, Washington DC, pp. 485–489 (2005)

Mobility, Data Mining and Privacy
the Experience of the GeoPKDD Project

Fosca Giannotti[*], Dino Pedreschi[**], and Franco Turini[***]

KDD Lab
ISTI-CNR, Pisa, and University of Pisa, Italy
Fosca.Giannotti@isti.cnr.it, Dino.Pedreschi@di.unipi.it,
Franco.Turini@di.unipi.it

1 Mobility Data

Our everyday actions, the way people live and move, leave digital traces in the information systems of the organizations that provide services through the wireless networks for mobile communication. The potential value of these traces in recording the human activities in a territory is becoming real, because of the increasing pervasiveness and positioning accuracy. The number of mobile phone users worldwide was recently estimated as 3 billion, i.e., one mobile phone every two people. On the other hand, the location technologies, such as GSM and UMTS, currently used by wireless phone operators are capable of providing an increasingly better estimate of a user's location, while the integration of various positioning technologies proceeds: GPS-equipped mobile devices can transmit their trajectories to some service provider (and the European satellite positioning system Galileo may improve precision and pervasiveness in the near future), Wi-Fi and Bluetooth devices may be a source of data for indoor positioning, Wi-Max can become an alternative for outdoor positioning, and so on. The consequence of this scenario, where communication and computing devices are ubiquitous and carried everywhere and always by people and vehicles, is that human activity in a territory may be *sensed* – not necessarily on purpose, but simply as a side effect of the ubiquitous services provided to mobile users. Thus, the wireless phone network, designed to provide mobile communication, can also be viewed as an infrastructure to gather mobility data, if used to record the location of its users at different times. The wireless networks, whose pervasiveness and localization precision increase while new location-based and context-based services are offered to mobile users, are becoming the nerves of our territory – in particular, our towns – capable of sensing and, possibly, recording our movements.

---

[*] Fosca Giannotti is a senior researcher at the Information Science and Technology Institute of the National Research Council in Pisa. She is the coordinator of GeoPKDD, and of the Italian National CNR project on Data mining, ontologies and semantic web.

[**] Dino Pedreschi is a full professor of computer science at the University of Pisa, and the coordinator of the work package *Privacy-aware spatio-temporal data mining* of the GeoPKDD project.

[***] Franco Turini is a full professor of computer science at the University of Pisa, and the coordinator of the *Privacy Observatory* of the GeoPKDD project.

F. Bonchi et al. (Eds.): PinkDD 2008, LNCS 5456, pp. 25–32, 2009.

From this perspective, we have today a chance of collecting and storing mobility data of unprecedented quantity, quality and timeliness at a very low cost: in principle, a dream for traffic engineers and urban planners, compelled until yesterday to gather data of limited size and precision only through highly expensive means such as field experiments, surveys to discover travelling habits of commuting workers and ad hoc sensors placed on streets.

However, there's a long way to go from mobility data to mobility knowledge. In the words of J.H. Poincaré, 'Science is built up with facts, as a house is with stones. But a collection of facts is no more a science than a heap of stones is a house.' Since databases became a mature technology and massive collection and storage of data became feasible at increasingly cheaper costs, a push emerged towards powerful methods for discovering knowledge from those data, capable of going beyond the limitations of traditional statistics, machine learning and database querying. This is what data mining is about.

2 Data Mining

Data mining is the process of automatically discovering useful information in large data repositories. Often, traditional data analysis tools and techniques cannot be used because of the massive volume of data gathered by automated collection tools, such as point-of-sale data, Web logs from e-commerce portals, earth observation data from satellites, genomic data. Sometimes, the non-traditional nature of the data implies that ordinary data analysis techniques are not applicable. The three most popular data mining techniques are predictive modeling, cluster analysis and association analysis.

-   In *predictive modeling*, the goal is to develop classification models, capable of predicting the value of a class label (or target variable) as a function of other variables (explanatory variables); the model is learnt from historical observations, where the class label of each sample is known: once constructed, a classification model is used to predict the class label of new samples whose class is unknown, as in forecasting whether a patient has a given disease based on the results of medical tests.
-   In *association analysis*, also called pattern discovery, the goal is precisely to discover patterns that describe strong correlations among features in the data or associations among features that occur frequently in the data. Often, the discovered patterns are presented in the form of association rules: useful applications of association analysis include market basket analysis, i.e. the task of finding items that are frequently purchased together, based on point-of-sale data collected at cash registers.
-   In *cluster analysis*, the goal is to partition a data set into groups of closely related data in such a way that the observations belonging to the same group, or cluster, are similar to each other, while the observations belonging to different clusters are not. Clustering can be used, for instance, to find segments of customers with a similar purchasing behavior or categories of documents pertaining to related topics.

Data mining is a step of *knowledge discovery in databases*, the so-called KDD process for converting raw data into useful knowledge. The KDD process consists of a series of transformation steps:

- *Data preprocessing*, which transforms the raw source data into an appropriate form for the subsequent analysis
- Actual *data mining*, which transforms the prepared data into patterns or models: classification models, clustering models, association patterns, etc.
- *Post processing* of data mining results, which assesses validity and usefulness of the extracted patterns and models, and presents interesting knowledge to the final users – business analysts, scientists, planners, etc. – by using appropriate visual metaphors or integrating knowledge into decision support systems.

Today, data mining is both a technology that blends data analysis methods with sophisticated algorithms for processing large data sets, and an active research field that aims at developing new data analysis methods for novel forms of data. One of the frontiers of data mining research, today, is precisely represented by spatiotemporal data, i.e., observations of events that occur in a given place at a certain time, such as the mobility data arriving from wireless networks.

3 Mobility Data Mining

Mobility data mining is, therefore, emerging as a novel area of research, aimed at the analysis of mobility data by means of appropriate patterns and models extracted by efficient algorithms; it also aims at creating a novel knowledge discovery process explicitly tailored to the analysis of mobility with reference to geography, at appropriate scales and granularity. In fact, movement always occurs in a given physical space, whose key semantic features are usually represented by geographical maps; as a consequence, the geographical background knowledge about a territory is always essential in understanding and analyzing mobility in such territory. Mobility data mining, therefore, is situated in a *Geographic Knowledge Discovery process* – a term first introduced by Han and Miller in [2] – capable of sustaining the entire chain of production from raw mobility data up to usable knowledge capable of supporting decision making in real applications.

As a prototypical example, assume that source data are positioning logs from mobile cellular phones, reporting user's locations with reference to the cells in the GSM network; these mobility data come as streams of raw log entries recording users entering a cell (*userID, time, cellID*). In this context, a geographic knowledge discovery process may be envisaged, composed of three main steps:

(1) *Trajectory reconstruction.* In this basic phase, the stream of raw mobility data has to be processed to obtain trajectories of individual moving objects; the resulting trajectories should be stored into appropriate repositories, such as a trajectory database or data warehouse.

(2) *Knowledge extraction.* Spatiotemporal data mining methods are needed to extract useful patterns and models out of trajectories:

- Trajectory clustering, the discovery of groups of 'similar' trajectories, together with a summary of each group;
- Trajectory patterns, the discovery of frequently followed (sub)trajectories;
- Trajectory classification, the discovery of decision rules, aimed at explaining the behaviour of current mobile users and predicting that of future users.

(3) *Knowledge delivery*. Extracted patterns are very seldom geographic knowledge prêt-a-porter: it is necessary to reason on patterns and on pertinent background knowledge, evaluate patterns' interestingness, refer them to geographic information, find out appropriate presentations and visual analytics methods.

We believe that if the geographic privacy-aware knowledge discovery process is built with solid theoretical foundations at an appropriate level of abstraction, it may become an enabling driver for knowledge-based applications of large impact, at different levels (societal, individual, business) in different challenging domains, such as in sustainable mobility, dynamic traffic monitoring, intelligent transportation systems and urban planning.

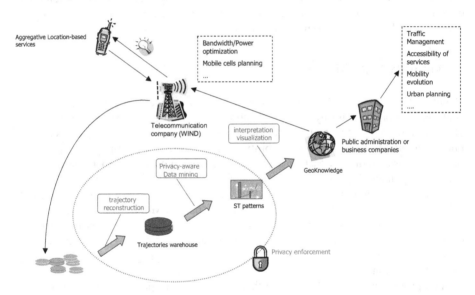

4 Achievements after Two Years of GeoPKDD

Toward its ambitious goal, the GeoPKDD project has achieved some key results in its first two years:

- Reconstruction of trajectories from streams of raw mobility data, and construction of a privacy-aware *trajectory warehouse* to store and analyze mobility data: a prototype trajectory warehouse has been created;
- *Spatio-temporal privacy-preserving data mining* and knowledge extraction algorithms, yielding mobility patterns and models: we designed, implemented and experimented a toolkit for spatio-temporal data mining for mobility data. Such

algorithms are scalable for massive datasets, and equipped with native methods for provably and measurably protecting privacy.

- *Geographic knowledge interpretation and visual analytics techniques* to deliver meaningful patterns to end users: we are constructing tools to assist the analytical process and the interpretation of the resulting mobility patterns, by means of: i) mechanisms for reasoning on the background geographic knowledge and on the extracted patterns; ii) mechanisms to express complex analytical queries into a data mining query language; ii) visual analytics methods: a suite of visually-driven tools for analyzing trajectory data has been developed and delivered. The toolkit incorporates a variety of data base processing, cluster analysis, and interactive visualization methods for the analysis of mobility data.

Two application demonstrators were selected in the industrial and societal domains, their feasibility assessed and their design developed: (1) the industrial application demonstrator is a set of analytical services for traffic managing in the city network of Milano; (2) the societal application demonstrator is an Educational Mobile Gaming service in the city of Amsterdam. The demonstrators are under development during the third, final year of the project.

4.1 Mining Useful Knowledge from Anonymous Mobility Data

Today, we are faced with the concrete possibility of pursuing an *archaeology of the present*: discovering from the digital traces of our mobile activity the knowledge that makes us comprehend timely and precisely the way we live, the way we use our time and our land. Thus, it is becoming possible, in principle, to understand how to live better by learning from our recent history, i.e. from the traces left behind us yesterday, or a few moments ago, recorded in the information systems and analysed to produce usable, timely and reliable knowledge. In simple words, we advocate that mobility data mining, defined as the collection and extraction of knowledge from mobility data, is the opportunity to construct novel services of great societal and economic impact.

However, there is a little path from opportunities to threats: We are aware that, on the basis of this scenario, there lies a flaw of potentially dramatic impact, namely the fact that the donors of the mobility data are the citizens, and making these data publicly available for the mentioned purposes would put at risk our own privacy, our natural right to keep secret the places we visit, the places we live or work at and the people we meet – all in all, the way we live as individuals. In other words, the personal mobility data, as gathered by the wireless networks, are extremely sensitive information; their disclosure may represent a brutal violation of the privacy protection rights, established in increasingly more laws and regulations internationally.

A genuine positivist researcher, with an unlimited trust in science and progress, may observe that, for the mobility-related analytical purposes, knowing the exact identity of individuals is not needed: anonymous data are enough to reconstruct aggregate movement behaviour, pertaining to whole groups of people, not to individual persons. This line of reasoning is also coherent with existing data protection regulations, such as that of the European Union, which states that personal data, once made anonymous, are not subject any longer to the restrictions of the privacy law. Unfortunately, this is not so easy: the problem is that anonymity means making reasonably impossible the re-identification, i.e. the linkage between the personal data of an individual and the identity of the individual itself. Therefore, transforming the data in such a way to guarantee anonymity is hard: supposedly anonymous data sets can leave unexpected doors open to malicious re-identification attacks, as some realistic examples show, in different domains such as medical patient data, Web search logs and location and trajectory data. Moreover, other possible breaches for privacy violation may be left open by the publication of the mining results, even in the case that the source data are kept secret by a trusted data custodian.

The bottom-line of this discussion is that protecting privacy when disclosing mobility knowledge is a non-trivial problem that, besides socially relevant, is scientifically attractive. As often happens in science, the problem is to find an optimal trade-off between two conflicting goals: from one side, we would like to have precise, fine-grained knowledge about mobility, which is useful for the analytic purposes; from the other side, we would like to have imprecise, coarse-grained knowledge about mobility, which puts us in repair from the attacks to our privacy. It is interesting that the same conflict – essentially between opportunities and risks – can be read either as a mathematical problem or as a social (or ethical or legal) challenge. Indeed, the privacy issues related to the ICTs can only be addressed through an alliance of technology, legal regulations and social norms. In the meanwhile, increasingly sophisticated privacy-preserving techniques are being studied. Their aim is to achieve appropriate levels of anonymity by means of controlled transformation of data and/or patterns – limited distortion that avoids the undesired side effect on privacy while preserving the possibility of discovering useful knowledge. A fascinating array of problems thus emerged, from the point of view of computer scientists and mathematicians, which already stimulated the production of important ideas and tools.

One direction of research is concerned with data anonymization, i.e., data transformations that guarantee a low, controlled probability of re-identification of personal data. In real life, people try to stay anonymous either by camouflage (i.e., pretending to be someone else or somewhere else) or by hiding in the crowd (i.e., pretending to be indistinguishable from a set of other people). Anonymity-preserving techniques have been developed for personal data following the two metaphors: transform the data using randomize noise (camouflage) and/or transform data using k-anonymity (generalize data to make each data subject indistinguishable from k others). Here, the challenge is to make sure that the transformed data can still be analyzed, yielding high quality patterns and models, comparable with those obtainable with the original data.

Another direction is secure multi-party data mining, where secure and efficient encryption techniques are used to compute a global data mining model from separate databases owned by different organizations, without the need of linking the separate databases together in a central server (a dangerous operation for privacy) and avoiding the risk of sharing (micro-) data among the different (possibly malicious) parties. In other words, the idea is to share the collective (aggregate) knowledge that can be mined from the collection of the separate databases, without actually sharing the separate data.

In the GeoPKDD project, we have developed techniques for anonymization and secure multi-party data mining for mobility data: we now have a rich software toolkit for

- *trajectory anonymization*, that combine k-anonymity and randomized perturbation to create high-quality anonymous versions of massive trajectory datasets, and

- *secure multi-party trajectory clustering*, that uses secret-sharing encryption techniques to create global clustering models from trajectory datasets partitioned among separate coopetative (cooperative – competitive) parties.

Hopefully, in the near future, it will be possible to reach a win–win situation: obtaining the advantages of collective mobility knowledge without divulging inadvertently any individual mobility knowledge. These results, if achieved, may have an impact on laws and jurisprudence, as well as on the social acceptance and dissemination of ubiquitous technologies.

4.2 Fostering the Interdisciplinary Dialogue through the GeoPKDD Privacy Observatory

Information and communication technologies (ICTs) touch on many aspects of our lives. The integration of ICTs is enhanced by the advent of mobile, wireless, and ubiquitous technologies. ICTs are increasingly embedded in common services, such as mobile and wireless communications, Internet browsing, credit card e-transactions, and electronic health records.

As ICT-based technologies become ubiquitous, our everyday actions leave behind increasingly detailed digital traces in the information systems of ICT-based service providers. For example, consumers of mobile phone technologies leave behind traces of geographic position to cellular provider records; Internet users leave behind traces the web page and packet requests of their computers in the access logs of domain and network administrators, and credit card transactions reveal the locations and times where purchases were completed.

As a result of the knowledge that may be discovered in the traces left behind by mobile users, the information systems of wireless networks pose potential opportunities for enhancing services, but threats abound. It should be noted that data, unto itself, is neither *good* nor *bad*. Rather, it is how the data is processed and applied, i.e., the purpose, that leads to a distinction between seemingly acceptable and unacceptable uses.

Protecting privacy when disclosing information is non-trivial. Anonymization and aggregation do not necessarily prevent attacks to privacy. For the same reason that the problem is scientifically attractive, it is socially relevant. It is interesting that the same conflict – essentially between opportunities and risks – can be either read as a mathematical problem or as a social (or ethical, or legal) challenge. Indeed, the privacy issues related to ICTs are unlikely to be solved by exclusively technological means: paraphrasing Rakesh Agrawal, one of the first researchers to address privacy issues in data management, any real solution to privacy problems can only be achieved through an alliance of technology, legal regulations and social norms.

It is exactly with this observation in mind that we created an observatory on privacy regulations within the GeoPKDD project.

5 Conclusion

Mobility, data mining and privacy: There is a new multi-disciplinary research frontier that is emerging at the crossroads of these three subjects, with plenty of challenging scientific problems to be solved and vast potential impact on real-life problems. Our GeoPKDD project is a key player in forging this research frontier, also thanks to the book produced by the GeoPKDD researchers, thoroughly aimed at substantiating the vision of convergence advocated above [1]. We gratefully acknowledge the commitment of the GeoPKDD consortium, which comprises:

- our KDD Lab., a joint research group of ISTI-CNR and University of Pisa,
- Hasselt University, Belgium
- EPFL - Ecole Polytechnique Fédérale de Lausanne, Switzerland
- Fraunhofer Institute for Intelligent Analysis and Information Systems, Germany
- Wageningen UR, Centre for GeoInformation, The Netherlands
- Polytechnic University of Madrid , Latingeo Laboratory, Spain
- Research Academic Computer Technology Institute, Athens, Greece
- Sabanci University, Istanbul, Turkey
- WIND Telecomunicazioni SpA, Italy
  together with the subcontractors:
- ICAR-CNR, Istituto di Calcolo e Reti ad Alte Prestazioni, Cosenza, Italy
- University of Milan, Italy
- University of Piraeus, Greece.

GeoPKDD is a project in the Future and Emerging Technologies programme of the Sixth Framework Programme for Research of the European Commission, FET-Open contract n: 014915.

References

1. Giannotti, F., Pedreschi, D. (eds.): Mobility, Data Mining and Privacy. Springer, Heidelberg (2008)
2. Miller, H.J., Han, J. (eds.): Geographic Data Mining and Knowledge Discovery. Taylor & Francis, Abington (2001)
3. Verykios, V.S., Damiani, M.L., Gkoulalas-Divanis, A.: Privacy and Security in Spatiotemporal Data and Trajectories. In: Pedreschi, D., Giannotti, F. (eds.) Mobility, Data Mining and Privacy, pp. 213–240. Springer, Heidelberg (2008)
4. Bonchi, F., et al.: Privacy in Spatiotemporal Data Mining. In: Pedreschi, D., Giannotti, F. (eds.) Mobility, Data Mining and Privacy, pp. 297–333. Springer, Heidelberg (2008)
5. Pedreschi, D., et al.: Privacy Protection: Regulations and Technologies, Opportunities and Threats. In: Pedreschi, D., Giannotti, F. (eds.) Mobility, Data Mining and Privacy, pp. 101–119. Springer, Heidelberg (2008)

Data and Structural k-Anonymity in Social Networks

Alina Campan and Traian Marius Truta

Department of Computer Science,
Northern Kentucky University,
Highland Heights, KY 41076, U.S.A.
{campana1,trutat1}@nku.edu

Abstract. The advent of social network sites in the last years seems to be a trend that will likely continue. What naive technology users may not realize is that the information they provide online is stored and may be used for various purposes. Researchers have pointed out for some time the privacy implications of massive data gathering, and effort has been made to protect the data from unauthorized disclosure. However, the data privacy research has mostly targeted traditional data models such as microdata. Recently, social network data has begun to be analyzed from a specific privacy perspective, one that considers, besides the attribute values that characterize the individual entities in the networks, their relationships with other entities. Our main contributions in this paper are a greedy algorithm for anonymizing a social network and a measure that quantifies the information loss in the anonymization process due to edge generalization.

Keywords: Privacy, Social Networks, K-Anonymity, Information Loss.

1 Introduction

While the ever increasing computational power, together with the huge amount of individual data collected daily by various agencies are of great value for our society, they also pose a significant threat to individual privacy. Datasets that store individual information have moved from simpler, traditional data models (such as microdata, where data is stored as one relational table, and each row represents an individual entity) to complex ones. The research in data privacy follows the same trend and tries to provide useful solutions for various data models. Although most of the privacy work has been done for healthcare data (usually in microdata form) mainly due to the Health Insurance Portability and Accountability Act regulation [11], privacy concerns have also been raised in other fields, where data usually takes a more complex form, such as location based services [3], genomic data [18], data streams [29], and social networks [9,10,15,31,32].

The advent of social networks in the last few years has accelerated the research in this field. Online social interaction has become very popular around the globe

F. Bonchi et al. (Eds.): PinkDD 2008, LNCS 5456, pp. 33–54, 2009.

and most sociologists agree that this trend will not fade away [27]. More and more social network datasets contain sensitive data. For example, epidemiology researchers are using social network datasets to study the relationship between sexual network structure and epidemic phase in sexually transmitted disease [21,28]. Other social networks datasets in areas such as e-mail communication [23] also benefit from privacy techniques tailored for social networks. Privacy in social networks is still in its infancy, and practical approaches are yet to be developed. A brief overview of proposed privacy techniques in social networks is given in the related work section.

We introduce in this paper a new anonymization approach for social network data that consists of nodes and relationships. A node represents an individual entity and is described by identifier (such as *Name* and *SSN*), quasi-identifier (such as *ZipCode* and *Sex*), and sensitive (such as *Diagnosis* and *Income*) attributes. A relationship is between two nodes and it is unlabeled, in other words, all relationships have the same meaning. To protect the social network data, we mask it according to the k-anonymity model (every node will be indistinguishable with at least other (k-1) nodes) [6,22,24], in terms of both nodes' attributes and nodes' associated structural information (neighborhood). Our anonymization method tries to disturb as little as possible the social network data, both the attribute data associated to the nodes, and the structural information. The method we use for anonymizing attribute data is generalization [22,25]. For structural anonymization we introduce a new method called edge generalization that does not insert into or remove edges from the social network dataset, similar to the one described in [31]. Although it incorporates a few ideas similar to those exposed in the related papers, our approach is new in several aspects. We embrace the k-anonymity model presented by Hay et al. [9,10], but we assume a much richer data model than just the structural information associated to the social network. We define an information loss measure that quantifies the amount of information loss caused by edge generalization (called *structural information loss*). We perform social network data clustering followed by anonymization through cluster collapsing. Our cluster formation process pays special attention to the nodes' attribute data and equally to the nodes' neighborhoods. This process can be user-balanced towards preserving more structural information of the network, as measured by the structural information loss, or the nodes' attribute values, which are quantified by the generalization information loss measure.

The remaining of this paper is structured as follows. Section 2 introduces our social network privacy model, in particular the concepts of edge generalization and k-anonymous masked social network. Section 3 starts by presenting the generalization and structural information loss measures, followed by our greedy social network anonymization algorithm. Section 4 contains comparative results, in terms of both generalization and structural information loss, for our algorithm and one of the existing privacy algorithms. Related work is presented in Section 5. The paper ends with future work directions and conclusions.

2   Social Network Privacy Model

We consider the social network modeled as a simple undirected graph $\mathcal{G} = (\mathcal{N}, \mathcal{E})$, where $\mathcal{N}$ is the set of nodes and $\mathcal{E} \subseteq \mathcal{N} \times \mathcal{N}$ is the set of edges. Each node represents an individual entity. Each edge represents a relationship between two entities.

The set of nodes, $\mathcal{N}$, is described by a set of attributes that are classified into the following three categories:

- $I_1, I_2, \ldots, I_m$ are *identifier* attributes such as *Name* and *SSN* that can be used to identify an entity.
- $Q_1, Q_2, \ldots, Q_q$ are *quasi-identifier* attributes such as *Zip_code* and *Sex* that may be known by an intruder.
- $S_1, S_2, \ldots, S_r$ are *confidential* or *sensitive* attributes such as *Diagnosis* and *Income* that are assumed to be unknown to an intruder.

We allow only binary relationships in our model. Moreover, we consider all relationships as being of the same type and, as a result, we represent them via unlabeled undirected edges. We also consider this type of relationship to be of the same nature as all the other "traditional" quasi-identifier attributes. We will refer to this type of relationship as *the quasi-identifier relationship*. In other words, the graph structure may be known to an intruder and used by matching it with known external structural information, therefore serving in privacy attacks that might lead to identity and/or attribute disclosure [12].

While the identifier attributes are removed from the published (masked) social network data, the quasi-identifier and the confidential attributes, as well as the graph structure, are usually released to the researchers/public. A general assumption, as noted, is that the values for the confidential attributes are not available from any external source. This assumption guarantees that an intruder cannot use the confidential attributes values to increase his/her chances of disclosure. Unfortunately, there are multiple techniques that an intruder can use to try to disclose confidential information. As pointed out in the microdata privacy literature, an intruder may use record linkage techniques between quasi-identifier attributes and external available information to glean the identity of individuals. Using the graph structure, an intruder is also able to identify individuals due to the uniqueness of the neighborhoods of various individuals. As shown in [9,10], when the structure of a random graph is known, the probability that there are two nodes with identical 3-radius neighborhoods is less than 2^{-cn}, where n represents the number of nodes in the graph, and c is a constant value, $c > 0$; this means that the vast majority of the nodes can be uniquely identified only on their 3-radius neighborhood structure.

A successful model for microdata privacy protection is k-anonymity, which ensures that every individual is indistinguishable with other $(k\text{-}1)$ individuals in terms of their quasi-identifier attributes' values [22,24]. For social network data, the k-anonymity model has to impose both the quasi-identifier attributes and the quasi-identifier relationship homogeneity, for groups of at least k individuals.

The generalization of the quasi-identifier attributes is one of the techniques widely used for microdata k-anonymization. It consists of replacing the actual value of an attribute with a less specific, more general value that is faithful to the original. We reuse this technique for the generalization of nodes attributes' values.

To our knowledge, the only method equivalent to our generalization of a quasi-identifier relationship that exists in the research literature appears in [31] and consists of collapsing clusters together with their component nodes' structure. Edge additions or deletions are currently used, in all the other approaches, to ensure nodes' indistinguishability in terms of their surrounding neighborhood; additions and deletions perturb to a large extent the graph structure and therefore they are not faithful to the original data. These methods are equivalent to randomization or perturbation techniques for microdata. We employ a generalization method for the quasi-identifier relationship similar to the one exposed in [31], but enriched with extra information, that will cause less damage to the graph structure, i.e. a smaller structural information loss.

Let n be the number of nodes from the set $\mathcal{N}$. Using a grouping strategy, one can partition the nodes from this set into v totally disjoint clusters: cl_1, cl_2, ..., cl_v. For simplicity we assume at this point that the nodes are not labeled (i.e., do not have attributes), and they can be distinguished only based on their relationships. Our goal is that any two nodes from any cluster to be also indistinguishable based on their relationships. To achieve this goal, we propose an edge generalization process, with two components: edge intra-cluster and edge inter-cluster generalization.

2.1 Edge Intra-cluster Generalization

Given a cluster cl, let $\mathcal{G}_{cl} = (cl, \mathcal{E}_{cl})$ be the subgraph of $\mathcal{G} = (\mathcal{N}, \mathcal{E})$ induced by cl. In the masked data, the cluster cl will be generalized to (collapsed into) a node, and the structural information we attach to it is the pair of values $(|cl|, |\mathcal{E}_{cl}|)$, where $|X|$ represents the cardinality of the set X. This information permits assessing some structural features about this region of the network that will be helpful in some applications. From the privacy standpoint, an original node within such a cluster is indistinguishable from the other nodes. At the same time, if more internal information was offered, such as the full nodes' connectivity inside a cluster, the possibility of disclosure would be too high, as discussed next.

When the cluster size is 2, the intra-cluster generalization doesn't eliminate any internal structural information, in other words the cluster's internal structure is fully recoverable from the masked information $(2, 0)$ or $(2, 1)$. For example, $(2, 0)$ means that the masked node represents two unconnected original nodes. Nevertheless, these two nodes are anyway indistinguishable from one another, inside the cluster, both in the presence and in the absence of an edge connecting them. This means that a required anonymity level 2 is achieved inside the cluster. However, when the number of nodes within a cluster is at least 3, it is possible to differentiate between various nodes if the cluster internal edges, $\mathcal{E}_{cl}$, are provided. Figure 1 shows comparatively several cases when the nodes can be distinguished and when they can be not (i.e., are anonymous) if the full

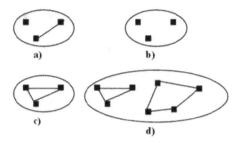

Fig. 1. 3-anonymous (b, c); non 3-anonymous (a); and non 7- anonymous (d)

internal structural information of the cluster was provided. It is easy to notice that a necessary condition that all nodes in a cluster must satisfy in order to be indistinguishable from each other is that all have the same degree. However, this condition is not sufficient, as shown in Figure 1.d, where all the nodes have a degree 2 and they can still be differentiated as belonging to one of the two cycles of the cluster. In this case, the anonymity level is 3, not 7.

2.2 Edge Inter-cluster Generalization

Given two clusters cl_1 and cl_2, let $\mathcal{E}_{cl_1,cl_2}$ be the set of edges having one end in each of the two clusters ($e \in \mathcal{E}_{cl_1,cl_2}$ iff $e \in \mathcal{E}$ and $e \in cl_1 \times cl_2$). In the masked data, this set of inter-cluster edges will be generalized to (collapsed into) a single edge and the structural information released for it is the value $|\mathcal{E}_{cl_1,cl_2}|$. This information permits assessing some structural features about this region of the network that might be helpful in some applications and it does not allow a presumptive intruder to differentiate between nodes within one cluster.

2.3 Masked Social Networks

Let's return to a fully specified social network and how to anonymize it. Given $\mathcal{G} = (\mathcal{N}, \mathcal{E})$, let X^i, $i = 1..n$, be the nodes in $\mathcal{N}$, where $n = |\mathcal{N}|$. We use the term *tuple* to refer only to the corresponding node attributes values (nodes' labels), without considering the relationships (edges) the node participates in. Also, we use the notation $X^i[C]$ to refer to the attribute C's value for the tuple X^i (the projection operation).

Once the nodes from $\mathcal{N}$ have been clustered into totally disjoint clusters $cl_1, cl_2, \ldots, cl_v$, in order to make all nodes in any cluster cl_i indistinguishable from one another in terms of their quasi-identifier attributes values, we generalize each cluster's tuples to the least general tuple that represents all tuples in that group.

There are several types of generalization available. Categorical attributes are usually generalized using generalization hierarchies, predefined by the data owner based on domain attribute characteristics (see Figure 2). For numerical attributes, generalization may be based on a predefined hierarchy or a hierarchy-free model. In our approach, for categorical attributes we use generalization

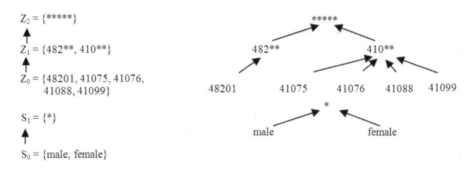

$Z_2 = \{*****\}$

$Z_1 = \{482**, 410**\}$

$Z_0 = \{48201, 41075, 41076,$
$\quad 41088, 41099\}$

$S_1 = \{*\}$

$S_0 = \{male, female\}$

Fig. 2. Domain and value generalization hierarchies for attributes *zip* and *gender*

based on predefined hierarchies at the cell level [16]. For numerical attributes we use the hierarchy-free generalization [13], which consists of replacing the set of values to be generalized with the smallest interval that includes all the initial values. We call generalization information for a cluster the minimal covering tuple for that cluster, and we define it as follows. (Of course, in this paragraph, generalization and coverage refer only to the quasi-identifier part of the tuples).

Definition 1. (generalization information of a cluster): Let $cl = \{X^1, X^2, \ldots, X^u\}$ be a cluster of tuples corresponding to nodes selected from $\mathcal{N}$, $\mathcal{QN} = \{N_1, N_2, \ldots, N_s\}$ be the set of numerical quasi-identifier attributes and $\mathcal{QC} = \{C_1, C_2, \ldots, C_t\}$ be the set of categorical quasi-identifier attributes. The **generalization information of** *cl* w.r.t. quasi-identifier attribute set $\mathcal{QI} = \mathcal{QN} \cup \mathcal{QC}$ is the "tuple" $gen(cl)$, having the scheme $\mathcal{QI}$, where:

- For each categorical attribute $C_j \in \mathcal{QI}$, $gen(cl)[C_j] = $ the lowest common ancestor in $\mathcal{H}_{C_j}$ of $\{X^1[C_j], \ldots, X^u[C_j]\}$. We denote by $\mathcal{H}_C$ the hierarchies (domain and value) associated to the categorical quasi-identifier attribute C.
- For each numerical attribute $N_j \in \mathcal{QI}$, $gen(cl)[N_j] = $ the interval $[min\{X^1[N_j], \ldots, X^u[N_j]\}, max\{X^1[N_j], \ldots, X^u[N_j]\}]$.

For a cluster *cl*, its generalization information $gen(cl)$ is the tuple having as value for each quasi-identifier attribute, numerical or categorical, the most specific common generalized value for all that attribute's values from *cl* tuples. In an anonymized graph, each tuple from cluster cl will have its quasi-identifier attributes values replaced by $gen(cl)$.

Given a partition of nodes for a social network $\mathcal{G}$, we are able to create an anonymized graph by using generalization information and edge intra-cluster generalization within each cluster and edge inter-cluster generalization between any two clusters.

Definition 2. (masked social network): Given an initial social network, modeled as a graph $\mathcal{G} = (\mathcal{N}, \mathcal{E})$, and a partition $\mathcal{S} = \{cl_1, cl_2, \ldots, cl_v\}$ of the nodes set $\mathcal{N}$, $\cup_{j=1}^{v} cl_j = \mathcal{N}$; $cl_i \cap cl_j = \emptyset$; $i, j = 1..v, i \neq j$; the corresponding **masked social network** $\mathcal{MG}$ is defined as $\mathcal{MG} = (\mathcal{MN}, \mathcal{ME})$, where:

- $\mathcal{MN} = \{Cl_1, Cl_2, \ldots, Cl_v\}$, Cl_j is a node corresponding to the cluster $cl_j \in \mathcal{S}$ and is described by the "tuple" $gen(cl_j)$ (the generalization information of cl_j, w.r.t. quasi-identifier attribute set) and the intra-cluster generalization pair $(|cl_j|, |E_{cl_j}|)$;
- $\mathcal{ME} \subseteq \mathcal{MN} \times \mathcal{MN}$; $(Cl_i, Cl_j) \in \mathcal{ME}$ iif $Cl_i, Cl_j \in \mathcal{MN}$ and $\exists X \in cl_i, Y \in cl_j$, such that $(X, Y) \in \mathcal{E}$. Each generalized edge $(Cl_i, Cl_j) \in \mathcal{ME}$ is labeled with the inter-cluster generalization value $|\mathcal{E}_{cl_i, cl_j}|$.

By construction, all nodes from a cluster cl collapsed into the generalized (masked) node Cl are indistinguishable from each other.

To have the k-anonymity property for a masked social network, we need to add one extra condition to Definition 2, namely that each cluster from the initial partition is of size at least k. The formal definition of a masked social network that is k-anonymous is presented below.

Definition 3. *(k-anonymous masked social network):* A masked social network $\mathcal{MG} = (\mathcal{MN}, \mathcal{ME})$, where $\mathcal{MN} = \{Cl_1, Cl_2, \ldots, Cl_v\}$, and $Cl_j = [gen(cl_j), (|cl_j|, |E_{cl_j}|)]$, $j = 1, \ldots, v$ is k-anonymous iff $|cl_j| \geq k$ for all $j = 1, \ldots, v$.

3   The *SaNGreeA* Algorithm

The algorithm described in this section, called the *SaNGreeA* (<u>S</u>ocial <u>N</u>etwork <u>Gree</u>dy <u>A</u>nonymization) algorithm, performs a greedy clustering processing to generate a k-anonymous masked social network, given an initial social network modeled as a graph $\mathcal{G} = (\mathcal{N}, \mathcal{E})$. Nodes from $\mathcal{N}$ are described by quasi-identifier and sensitive attributes and edges from $\mathcal{E}$ are undirected and unlabeled.

First, the algorithm establishes a "good" partitioning of all nodes from $\mathcal{N}$ into clusters. Next, all nodes within each cluster are made uniform with respect to the quasi-identifier attributes and the quasi-identifier relationship. This homogenization is achieved by using generalization, both for the quasi-identifier attributes and the quasi-identifier relationship, as explained in the previous section.

But how is the clustering process conducted such that a good partitioning is created and what does "good" mean? In order for the requirements of the k-anonymity model to be fulfilled, each cluster has to contain at least k tuples. Consequently, a first criterion to lead the clustering process is to ensure that each cluster has enough elements. As it is well-known, (attribute and relationship) generalization results in information loss. Therefore, a second criterion used during clustering is to minimize the information lost between the initial social network data and its masked version, caused by the subsequent cluster-level quasi-identifier attributes and relationship generalization. In order to obtain good quality masked data, and also to permit the user to control the type and the quantity of information loss he/she can afford, the clustering algorithm uses two information loss measures. One quantifies how much *descriptive* data detail is lost through quasi-identifier attributes generalization - we call this metric the generalization information loss measure. The second measure quantifies how much *structural* detail is lost through the quasi-identifier relationship generalization and it is called structural information loss. In the remainder of this section, these two information loss measures and the *SaNGreeA* algorithm are introduced.

3.1   Generalization Information Loss

The generalization of quasi-identifier attributes reduces the quality of the data. To measure the amount of information loss, several cost measures were introduced [4,7,13]. In our social network privacy model, we use the generalization information loss measure as introduced and described in [4]:

Definition 4. *(generalization information loss):* Let cl be a cluster, $gen(cl)$ its generalization information, and $\mathcal{QI} = \{N_1, N_2, \ldots, N_s, C_1, C_2, \ldots, C_t\}$ the set of quasi-identifier attributes. The ***generalization information loss*** caused by generalizing quasi-identifier attributes of the cl tuples to $gen(cl)$ is:

$$GIL(cl) = |cl| \cdot \left(\sum_{j=1}^{s} \frac{size(gen(cl)[N_j])}{size(min_{X \in \mathcal{N}}(X[N_j]), max_{X \in \mathcal{N}}(X[N_j]))} + \right.$$

$$\left. \sum_{j=1}^{t} \frac{height(\Lambda(gen(cl)[C_j]))}{height(\mathcal{H}_{C_j})} \right),$$

where:

- $|cl|$ denotes the cluster cl's cardinality;
- $size([i_1, i_2])$ is the size of the interval $[i_1, i_2]$, i.e., $(i_2 - i_1)$;
- $\Lambda(w)$, $w \in \mathcal{H}_{C_j}$ is the subhierarchy of $\mathcal{H}_{C_j}$ rooted in w;
- $height(\mathcal{H}_{C_j})$ denotes the height of the tree hierarchy $\mathcal{H}_{C_j}$.

Definition 5. *(total generalization information loss):* Total generalization information loss produced when masking the graph $\mathcal{G}$ based on the partition $\mathcal{S} = \{cl_1, cl_2, \ldots, cl_v\}$, denoted by $GIL(\mathcal{G}, \mathcal{S})$, is the sum of the generalization information loss measure for each of the clusters in $\mathcal{S}$:

$$GIL(\mathcal{G}, \mathcal{S}) = \sum_{j=1}^{v} GIL(cl_j).$$

In the above measures, the information loss caused by the generalization of each quasi-identifier attribute value, for any tuple, is a value between 0 and 1. This means that each tuple contributes to the total generalization loss with a value between 0 and $(s + t)$ (the number of quasi-identifier attributes). Since the graph has n tuples, the total generalization information loss is a number between 0 and $n \cdot (s + t)$. To be able to compare this measure with the structural information loss, we chose to normalize both of them to the range $[0, 1]$.

Definition 6. *(normalized generalization information loss):* The ***normalized generalization information loss*** obtained when masking the graph $\mathcal{G}$ based on the partition $\mathcal{S} = \{cl_1, cl_2, \ldots, cl_v\}$, denoted by $NGIL(\mathcal{G}, \mathcal{S})$, is the sum of the generalization information loss measure for each of the clusters in $\mathcal{S}$:

$$NGIL(\mathcal{G}, \mathcal{S}) = \frac{GIL(\mathcal{G}, \mathcal{S})}{n \cdot (s + t)}.$$

3.2 Structural Information Loss

We introduce next a measure to quantify the structural information which is lost when anonymizing a graph through collapsing clusters into nodes, together with their neighborhoods.

Information loss in this case quantifies the probability of error when trying to reconstruct the structure of the initial social network from its masked version. There are two components for the structural information loss: the *intra-cluster structural loss* and the *inter-cluster structural loss* components.

Let cl be a cluster of nodes from $\mathcal{N}$, and $\mathcal{G}_{cl} = (cl, \mathcal{E}_{cl})$ be the subgraph induced by cl in $\mathcal{G} = (\mathcal{N}, \mathcal{E})$. When cl is replaced (collapsed) in the masked graph $\mathcal{MG}$ with the node Cl described by the pair $(|cl|, |\mathcal{E}_{cl}|)$, the probability of an edge to exist between any pair of nodes from cl is $|\mathcal{E}_{cl}|/ \binom{|cl|}{2}$. Therefore, for each of the real edges from cluster cl, the probability that someone wrongly labels it as a non-edge is $1 - |\mathcal{E}_{cl}|/ \binom{|cl|}{2}$. At the same time, for each pair of unconnected edges from cluster cl, the probability that someone wrongly labels it as an edge is $|\mathcal{E}_{cl}|/ \binom{|cl|}{2}$.

Definition 7. *(intra-cluster structural information loss):* The *intra-cluster structural information loss* $(intraSIL)$ is the probability of wrongly labeling a pair of nodes in cl as an edge or as an unconnected pair. As there are $|\mathcal{E}_{cl}|$ edges, and $\binom{|cl|}{2} - \mathcal{E}_{cl}$ pairs of unconnected nodes in cl,

$$intraSIL(cl) = \left(\left(\binom{|cl|}{2} - |\mathcal{E}_{cl}| \right) \cdot |\mathcal{E}_{cl}|/ \binom{|cl|}{2} + |\mathcal{E}_{cl}| \cdot \left(1 - |\mathcal{E}_{cl}|/ \binom{|cl|}{2} \right) \right) =$$
$$2 \cdot |\mathcal{E}_{cl}| \cdot \left(1 - |\mathcal{E}_{cl}|/ \binom{|cl|}{2} \right).$$

Reasoning in the same manner as above, we introduce the second structural information loss measure.

Definition 8. *(inter-cluster structural information loss):* The *inter-cluster structural information loss* $(interSIL)$) is the probability of wrongly labeling a pair of nodes (X, Y), where $X \in cl_1$ and $Y \in cl_2$, as an edge or as an unconnected pair. As there are $|\mathcal{E}_{cl_1, cl_2}|$ edges, and $|cl_1| \cdot |cl_2| - |\mathcal{E}_{cl_1, cl_2}|$ pairs of unconnected nodes between cl_1 and cl_2,

$$interSIL(cl_1, cl_2) = (|cl_1| \cdot |cl_2| - |\mathcal{E}_{cl_1, cl_2}|) \cdot \frac{|\mathcal{E}_{cl_1, cl_2}|}{|cl_1| \cdot |cl_2|} + |\mathcal{E}_{cl_1, cl_2}| \cdot \left(1 - \frac{|\mathcal{E}_{cl_1, cl_2}|}{|cl_1| \cdot |cl_2|} \right)$$
$$= 2 \cdot |\mathcal{E}_{cl_1, cl_2}| \cdot \left(1 - \frac{|\mathcal{E}_{cl_1, cl_2}|}{|cl_1| \cdot |cl_2|} \right).$$

Now, we have all the tools to introduce the total structural information loss measure.

Definition 9. *(total structural information loss):* The **total structural information loss** obtained when masking the graph $\mathcal{G}$ based on the partition $S = \{cl_1, cl_2, \ldots, cl_v\}$, denoted by $SIL(\mathcal{G}, S)$, is the sum of all inter-cluster and intra-cluster structural information loss values:

$$SIL(\mathcal{G}, S) = \sum_{j=1}^{v}(intraSIL(cl_j)) + \sum_{i=1}^{v}\sum_{j=i+1}^{v}(interSIL(cl_i, cl_j)).$$

We analyze the $intraSIL(cl)$ function for a given fixed cluster cl and a variable number of edges in the cluster, $|\mathcal{E}_c l|$, in other words, we consider $intraSIL(cl)$ a function of a variable $|\mathcal{E}_c l|$. Based on Definition 7, this function is (we use f to denote the function and x the variable number of edges):

$$f : \left\{0, 1, \ldots, \binom{|cl|}{2}\right\} \to \Re,$$

$$f(x) = 2 \cdot x \cdot \left(1 - x / \binom{|cl|}{2}\right).$$

Using the first and second derivative function it can easily be determined that the maximum value the function f takes is for

$$x = \binom{|cl|}{2} / 2 = \frac{|cl| \cdot (|cl| - 1)}{4}.$$

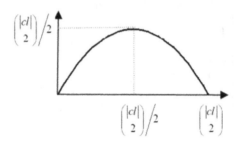

Fig. 3. *intraSIL* as a function of number of edges for $|cl|$ fixed

Figure 3 shows the graphical representation of the $f(x)$ function. As it can be seen, the smallest values of the function correspond to clusters that are either unconnected graphs (no edges) or completely connected graphs. The maximum function value corresponds to a cluster that has the number of edges equal to half of the number of all the pairs of nodes in the cluster.

A similar analysis, with the same results, can be conducted for the function $interSIL(cl_1, cl_2)$, seen as a function of one variable $|\mathcal{E}_{cl_1, cl_2}|$, when clusters cl_1 and cl_2 are fixed. This function has a similar behavior with $intraSIL(cl)$. Namely, minimum is reached when $|\mathcal{E}_{cl_1, cl_2}|$ is either 0 or the maximum possible value $|cl_1| \cdot |cl_2|$, and the maximum is reached when $|\mathcal{E}_{cl_1, cl_2}|$ is equal to $|cl_1| \cdot |cl_2|/2$.

This analysis suggests that a smaller structural information loss corresponds to clusters in which nodes have similar connectivity properties with one another or, in other words, when cluster's nodes are either all connected (or unconnected) among them and with the nodes in other clusters. We will use this result in our anonymization algorithm.

To normalize the structural information loss, we compute the maximum values for $intraSIL(cl)$ and $interSIL(cl_1, cl_2)$. As illustrated in Figure 3, the maximum value for $intraSIL(cl)$ is $|cl| \cdot (|cl| - 1)/4$. Similarly, the maximum value for $interSIL(cl_1, cl_2)$ is $|cl_1| \cdot |cl_2|/2$. Using Definition 9, we derive the maximum total structural information loss value as:

$$\sum_{j=1}^{v} \frac{|cl_j| \cdot (|cl_j| - 1)}{4} + \sum_{i=1}^{v} \sum_{j=i+1}^{v} \frac{|cl_i| \cdot |cl_j|}{4} =$$
$$\tfrac{1}{4} \cdot \left(\sum_{j=1}^{v} |cl_j|^2 + 2 \cdot \sum_{i=1}^{v} \sum_{j=i+1}^{v} |cl_i| \cdot |cl_j| \right) - \tfrac{1}{4} \sum_{j=1}^{v} |cl_j| =$$
$$\tfrac{1}{4} \left(\sum_{j=1}^{v} |cl_j| \right)^2 - \tfrac{1}{4} \sum_{j=1}^{v} |cl_j| = \frac{n \cdot (n-1)}{4}.$$

The minimum total structural information loss is 0, and it is obtained for a graph with no edges or for a complete graph.

Definition 10. *(normalized structural information loss):* The **normalized structural information loss** obtained when masking the graph $\mathcal{G}$ with n nodes, based on the partition $\mathcal{S} = \{cl_1, cl_2, \ldots, cl_v\}$, denoted by $NSIL(\mathcal{G}, \mathcal{S})$, is:

$$NSIL(\mathcal{G}, \mathcal{S}) = \frac{SIL(\mathcal{G}, \mathcal{S})}{(n \cdot (n-1)/4)}.$$

The normalized structural information loss is in the range $[0, 1]$.

3.3   The Anonymization Algorithm

The *SaNGreeA* algorithm puts together in clusters nodes that are as similar as possible, both in terms of their quasi-identifier attribute values, and in terms of their neighborhood structure. This greedy approach tries to minimize the generalization information loss and the structural information loss for the generated k-anonymous masked social network.

To assess the proximity between nodes with respect to quasi-identifier attributes, we use the normalized generalization information loss. However, the structural information loss cannot be computed during the clusters creation process, as long as the entire partitioning is not known. Therefore, we chose to guide the clustering process using a different measure. This measure quantifies the extent in which the neighborhoods of two nodes are similar with each other, i.e., the nodes present the same connectivity properties, or are connected / disconnected among them and with others in the same way.

To assess the proximity of two nodes' neighborhoods, we proceed as follows. Given $\mathcal{G} = (\mathcal{N}, \mathcal{E})$, assume that nodes in $\mathcal{N}$ have a particular order,

$\mathcal{N} = \{X^1, X^2, \ldots, X^n\}$. The neighborhood of each node X^i can be represented as an n-dimensional boolean vector $B_i = (b_1^i, b_2^i, \ldots, b_n^i)$, where the jth component of this vector, b_j^i, is 1 if there is an edge $(X^i, X^j) \in \mathcal{E}$, and 0 otherwise, $\forall j = 1..n; j \neq i$. We consider the value b_i^i to be *undefined*, and therefore not equal with 0 or 1. We use a classical distance measure for this type of vector, the *symmetric binary distance* [8].

Definition 11. *(distance between two nodes):* The **distance between two nodes** $(X^i$ and $X^j)$ described by their associated n-dimensional boolean vectors B_i and B_j is:

$$dist(X^i, X^j) = \frac{|\{\ell | \ell=1..n \wedge \ell \neq i,j; b_\ell^i \neq b_\ell^j\}|}{n-2}.$$

We exclude from the two vectors comparison their elements i and j, which are undefined for X^i and respectively for X^j. As a result, the total number of elements compared is reduced by 2.

In the cluster formation process, our greedy approach will select a node to be added to an existing cluster. To assess the structural distance between a node and a cluster we use the following measure.

Definition 12. *(distance between a node and a cluster):* The **distance between a node** X **and a cluster** cl is defined as the average distance between X and every node from cl:

$$dist(X, cl) = \frac{\sum_{X^j \in cl} dist(X, X^j)}{|cl|}.$$

We note that both distance measures take values between 0 and 1, and they can be used in the cluster formation process in combination with the normalized generalization information loss.

Although this is not formally proved, but shown to be effective in our experiments, by putting together in clusters nodes that are the closest according to the average distance measure, the *SaNGreeA* algorithm will produce a good masked network, with a small structural information loss.

Using the above introduced measures, we explain how clustering is performed for a given initial social network $\mathcal{G} = (\mathcal{N}, \mathcal{E})$. The clusters are created one at a time. To form a new cluster, a node in $\mathcal{N}$ with the maximum degree and not yet allocated to any cluster is selected as a seed for the new cluster. Then the algorithm gathers nodes to this currently processed cluster until it reaches the desired cardinality k. At each step, the current cluster grows with one node. The selected node has to be unallocated yet to any cluster and to minimize the cluster's information loss growth, quantified as a weighted measure that combines $NGIL$ and $dist$. The parameters α and β, with $\alpha + \beta = 1$, control the relative importance given to the total generalization information loss (the parameter α) and the total structural information loss (the parameter β) and are user-defined.

It is possible, when n is not a multiple of k, that the last constructed cluster will contain less than k nodes. In that case, this cluster needs to be dispersed between the previously constructed groups. Each of its nodes will be added to the cluster whose information loss will minimally increase by that node addition.

The pseudocode for our social network anonymization algorithm is shown next.

```
Algorithm SaNGreeA is

Input  G = (N, E) - a social network
       k - as in k-anonymity
       α and β, α + β = 1 - user-defined weight parameters;
       allow controlling the balancing between GIL and SIL.

Output S = {cl₁, cl₂, ..., cl_v}; ∪ᵥⱼ₌₁ clⱼ = N; cl_i ∩ cl_j = ∅,
       i, j = 1..v, i ≠ j; |cl_j| ≥ k, j = 1..v - a set of clusters
       that ensures k-anonymity for MG = (MN, ME) so that
       a cost measure is optimized;
S = ∅;
i = 1;
Repeat
       X^seed = a node with maximum degree from N;
       cl_i = {X^seed};
       // N keeps track of nodes not yet distributed to clusters
       N = N - {X^seed};
       Repeat
           X* = argmin_{X∈N}(α · NGIL(G₁, S₁) + β · dist(X, cl_i));
           // X* is the node within N (unselected nodes) that
           // produces the minimal information loss growth when
           // added to cl_i
           // G₁ - the subgraph induced by cl ∪ {X*} in G;
           // S₁ - a partition with one cluster cl ∪ {X*}
           cl_i = cl_i ∪ {X*};
           N = N - {X*};
       Until (cl_i has k elements) or (N == ∅);
       If (|cl_i| < k) then
           DisperseCluster(S, cl_i); // only for the last cluster
       Else
           S = S ∪ {cl_i};
           i++;
       End If;
Until N == ∅;
End SaNGreeA.

Function DisperseCluster(S, cl)
       For every X ∈ cl do
           cl_u = FindBestCluster(X, S);
```

$$cl_u = cl_u \cup \{X\};$$
    End For;
End *DisperseCluster*;

Function *FindBestCluster* (X, S) is
    bestCluster = null;
    infoLoss = ∞;
    For every $cl_j \in S$ do
        If $\alpha \cdot NGIL(\mathcal{G}_1, \mathcal{S}_1) + \beta \cdot dist(X, cl_i) <$ infoLoss then
            infoLoss = $\alpha \cdot NGIL(\mathcal{G}_1, \mathcal{S}_1) + \beta \cdot dist(X, cl_i)$;
            bestCluster = cl_j;
        End If;
    End For;
    Return bestCluster;
End *FindBestCluster*;

Because *SaNGreeA* is a greedy algorithm, that selects a solution from the search space (i.e., the set of all partitions of $\mathcal{N}$ consisting of subsets of k or more nodes) based on local optima of the two criterion measures, the algorithm will find a good solution to the anonymization problem, but not the best existing solution. The time complexity of *SaNGreeA* is $O(n^2)$. However, an efficient (sub-exponential) method to find the optimal solution is not known: the k-anonymization for microdata has been proved to be NP-hard [19] and our optimization problem for social network data is similar, with the only difference of having to minimize two measures of the amount of information in the initial data that is not released.

We show next an example that illustrates the concepts of generalization and structural information loss as well as how the obtained solution is dependent of the selection of α and β.

Suppose the social network $\mathcal{G}_{ex}$ depicted in Figure 4 is given. It contains nine nodes, described by the quasi-identifier attributes *age, zip* and *gender*. The *age* quasi-identifier is numerical, *zip* and *gender* are categorical - their predefined domain and value generalization hierarchies are presented in Figure 2. The quasi-identifier attributes' values for all nodes are depicted in Table 1.

By running the *SaNGreeA* algorithm for this set of data for ($k = 3$, $\alpha = 1$, and $\beta = 0$) and ($k = 3$, $\alpha = 0$, and $\beta = 1$) respectively, we obtain the 3-anonymous masked social networks $\mathcal{MG}_{e1}$ and $\mathcal{MG}_{e2}$ depicted in Figure 5. We did not show in the figure the generalization information for the clusters, but this can be easily computed; for instance, $gen(cl_2) = \{[25 - 27], 410 * *, male\}$.

In Table 2 we show the information loss measures' values computed based on Definitions 4 - 10. As expected, due to the weights choice, $\mathcal{MG}_{e1}$ is a better solution in terms of total generalization information loss than $\mathcal{MG}_{e2}$ and $\mathcal{MG}_{e2}$ outperforms $\mathcal{MG}_{e1}$ with respect to total structural information loss.

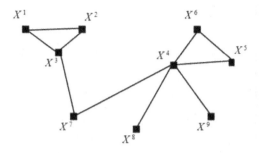

Fig. 4. The Social Network $\mathcal{G}_{ex}$

Table 1. The quasi-identifier attributes' values for $\mathcal{G}_{ex}$ nodes

Node	age	zip	gender
X^1	25	41076	male
X^2	25	41075	male
X^3	27	41076	male
X^4	35	41099	male
X^5	38	48201	female
X^6	36	41075	female
X^7	30	41099	male
X^8	28	41099	male
X^9	33	41075	female

4   Experimental Results

In this section we compare the *SaNGreeA* algorithm and the anonymization algorithm proposed in [31], which is based on collapsing clusters as formed by any classical k-anonymization algorithm for microdata [4,13]. For our experiments, we use the clustering algorithm introduced in [4]. Comparisons of *SaNGreeA* with other existing algorithms for anonymizing social networks [2,9,32] are not feasible, as those algorithms do not take into consideration a full range of quasi-identifier attributes, as we do; usually they consider at most one quasi-identifier attribute and, of course, the quasi-identifier relationship. Another difference that impeded comparison with other algorithms is the incompatibility in how relationships are seen across different anonymization approaches: single type versus multiple types of relationships, relationships with or without attributes etc. Zheleva's algorithm seems to be the only compatible and obviously comparable with ours.

The comparison we present between the *SaNGreeA* algorithm and the Zheleva's algorithm [31] is made with respect to the quality of the results they produce, measured against the normalized generalization information loss and the normalized structural information loss.

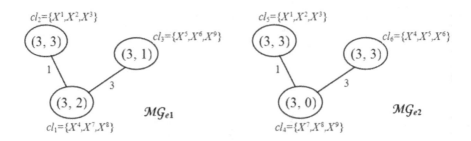

Fig. 5. The k-anonymous masked social networks $\mathcal{MG}_{e1}$ and $\mathcal{MG}_{e2}$

Table 2. Information loss values

$(\mathcal{G}, \mathcal{MG})$	$(\mathcal{G}_{ex}, \mathcal{MG}_{e1})$ with partition $\mathcal{S}_1 = \{\{X^4, X^7, X^8\},$ $\{X^1, X^2, X^3\},$ $\{X^5, X^6, X^9\}\}$	$(\mathcal{G}_{ex}, \mathcal{MG}_{e2})$ with partition $\mathcal{S}_2 = \{\{X^4, X^5, X^6\},$ $\{X^1, X^2, X^3\},$ $\{X^7, X^8, X^9\}\}$
$GIL, NGIL$	$GIL(\mathcal{G}, \mathcal{S}_1) =$ $3 \cdot \left(\frac{7}{13} + 0 + 0\right) + 3 \cdot \left(\frac{2}{13} + \frac{1}{2} + 0\right)$ $+ 3 \cdot \left(\frac{5}{13} + 1 + 0\right) = 7.730$ $NGIL(\mathcal{G}, \mathcal{S}_1) = \frac{7.730}{9 \cdot 3} = 0.286$	$GIL(\mathcal{G}, \mathcal{S}_2) =$ $3 \cdot \left(\frac{3}{13} + 1 + 1\right) + 3 \cdot \left(\frac{2}{13} + \frac{1}{2} + 0\right)$ $+ 3 \cdot \left(\frac{5}{13} + \frac{1}{2} + 1\right) = 14.307$ $NGIL(\mathcal{G}, \mathcal{S}_2) = \frac{14.307}{9 \cdot 3} = 0.529$
$intraSIL$	$intraSIL(cl_1) = \frac{4}{3}$ $intraSIL(cl_2) = 0$ $intraSIL(cl_3) = \frac{4}{3}$	$intraSIL(cl_4) = 0$ $intraSIL(cl_5) = 0$ $intraSIL(cl_6) = 0$
$interSIL$	$interSIL(cl_1, cl_2) = \frac{16}{9}$ $interSIL(cl_1, cl_3) = 4$ $interSIL(cl_2, cl_3) = 0$	$interSIL(cl_4, cl_5) = \frac{16}{9}$ $interSIL(cl_4, cl_6) = 4$ $interSIL(cl_5, cl_6) = 0$
$SIL, NSIL$	$SIL(\mathcal{G}, \mathcal{S}_1) = 8.444$ $NSIL(\mathcal{G}, \mathcal{S}_1) = 0.469$	$SIL(\mathcal{G}, \mathcal{S}_2) = 5.777$ $NSIL(\mathcal{G}, \mathcal{S}_2) = 0.320$

The two algorithms were implemented in Java; tests were executed on a dual CPU machine with 3.00GHz and 4GB of RAM, running Windows NT Professional. Experiments were performed for a social network with 300 nodes randomly selected from the Adult dataset from the UC Irvine Machine Learning Repository [20]; we refer to this set as $\mathcal{N}$.

In all the experiments, we considered a set of six quasi-identifier attributes: *age*, *workclass*, *marital-status*, *race*, *sex*, and *native-country*. The *age* attribute was the only numerical quasi-identifier, the other five attributes are categorical. Figure 6 depicts the generalization hierarchy for the *native-country* attribute, the categorical attribute with the most developed hierarchy. The remaining four quasi-identifier categorical attributes have the following heights for their corresponding value generalization hierarchies: *workclass* - 1, *marital-status* - 2, *race* - 1, and *sex* - 1. As already explained, for the quasi-identifier numerical attribute we used hierarchy-free generalization [13].

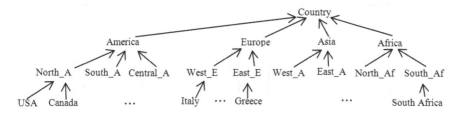

Fig. 6. The value hierarchy for the quasi-identifier attribute *native-country*

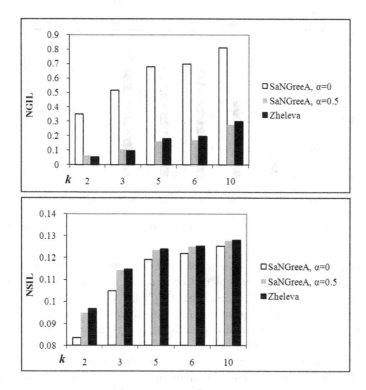

Fig. 7. *NGIL* and *NSIL* for *Random Graph*

Three different synthetic sets of edges were considered, all generated using *GTGraph*, a synthetic graph generator suite [1]. The first edge set corresponds to a random graph with an average vertex degree of 10; we refer to this edge set as $\mathcal{E}_1$. For producing $\mathcal{E}_1$, we used the random graph generator included in the *GTGraph* suite and we replaced with other random edges all but one of the multiple edges between the same pair of vertices. The second edge set we experimented with was generated in agreement with the power law distribution and the small-world characteristic, which are the two most important properties for many real-world social networks [32]; we refer to this edge set as $\mathcal{E}_2$. For

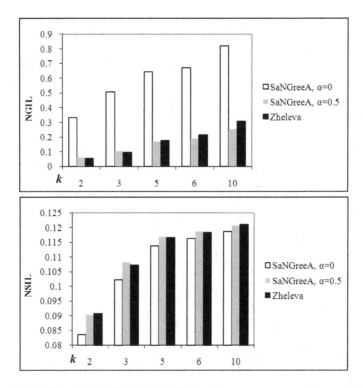

Fig. 8. *NGIL* and *NSIL* for R_MAT Graph, average vertex degree of 9.52

producing $\mathcal{E}_2$, we used the R_MAT graph model [5] and generator included in the *GTGraph* suite. We randomly replaced or removed the multiple edges between the same pair of vertices. The resulting graph $(\mathcal{N}, \mathcal{E}_2)$ had an average vertex degree of 9.52. The third edge set we experimented with was similar with the second one, in the sense that it was generated in agreement with the power law distribution and the small-world characteristic; it differed from the second one on the average vertex degree, which was 5. We refer to this edge set as $\mathcal{E}_3$.

The *SaNGreeA* algorithm and the algorithm introduced in [31] were applied to these three social networks, $\mathcal{G}_1 = (\mathcal{N}, \mathcal{E}_1)$, $\mathcal{G}_2 = (\mathcal{N}, \mathcal{E}_2)$, and $\mathcal{G}_3 = (\mathcal{N}, \mathcal{E}_3)$, for different k values, $k = 2, 3, 5, 6$, and 10. Figures 7, 8, and 9 present comparatively the normalized generalization information loss and the normalized structural information loss values of the results produced by applying the two algorithms, for the graphs $\mathcal{G}_1$, $\mathcal{G}_2$, and respectively $\mathcal{G}_3$, for all considered k values, and for two different value sets for the parameters α and β in the *SaNGreeA* algorithm. The (α, β) occurrences we used are $(0.0, 1.0)$ and respectively $(0.5, 0.5)$. The pair $(0.0, 1.0)$ guides the algorithm towards minimizing the structural information loss, without giving any consideration to the generalization information loss factor. The pair $(0.5, 0.5)$ signifies a request for the algorithm to equally weight both information loss components in the cluster formation process. As expected, while both tested algorithms, with all different parameters selections, produce a

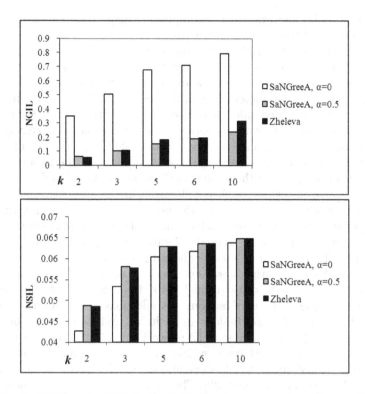

Fig. 9. $NGIL$ and $NSIL$ for R_MAT Graph, average vertex degree of 5

k-anonymized masked social network, the data utility conserved by each solution is different. For the *SaNGreeA* experiments the structural information loss is, in general, smaller than in the Zheleva's algorithm case.

This comes with the cost of greater generalization information loss. Since it is based on defining the weight of generalization/structural information loss, our algorithm is very flexible and allows the user to customize the amount of generalization and/or structural information loss he agrees to in a particular anonymization task. A special note is worth to be made. Our algorithm can be tuned to be equivalent to Zheleva's (when the last one bases its cluster formation on the greedy algorithm explained in [4]), by appropriately setting (α, β) parameters to $(1.0, 0.0)$. The general rule is to set β to a value greater than α's when more structural information needs to be preserved when anonymizing the network; and vice versa, α has to be set to a value greater than β's when more generalization information needs to be preserved.

5   Related Work

The research in social networks privacy is very recent, and many questions are still to be answered. Only a few researchers have explored this integrative field of privacy in social networks from a computing perspective. We briefly present a short overview of the approaches we are aware of.

Zheleva and Geetor consider the problem where relationships between different individual entities in a network must be protected, and they called this problem link re-identification [31]. Their anonymization approach functions in two steps: first anonymize descriptive data from the graph nodes (the individual entities) to achieve k-anonymity or t-closeness [14], without considering in this step, in any way, the relationships between the network nodes. Their next step is to anonymize the network's structure, by controlled edge removal, in different flavors, each with different success likelihood: edges can all be removed, only a user-specified percentage of them, none of them, or can be generalized at a cluster level. Our work is closest to theirs. However, in our approach we anonymize the social network data at once, i.e., the nodes and edges anonymizations are integrated together in our masking algorithm and occur concurrently.

Other researchers have focused on developing a concept similar to k-anonymity for graph data. Hay et al. defines k-candidate anonymity based on the similarity of neighborhoods, in other words every node has at least k candidate nodes from which it is hard to be distinguished [9,10]. In order to satisfy this property, the graph data suffers a series of random edge additions and deletions. The nodes also do not contain attributes besides an identifier, and the edges are of a single type. Zhou and Pei have a similar social network model, they consider the nodes to be labeled (having one attribute, which can be seen as a quasi-identifier) and that only the near vicinity (1-radius neighborhood) of some target individuals is completely known to an intruder [32]. Their solution generalizes the node labels (attribute values) and adds extra edges to create similar neighborhoods. Their approach guarantees that an adversary with the knowledge of a 1-neigborhood cannot identify any individual with a confidence higher than $1/k$. Liu and Terzi introduced the concept of k-degree anonymous graph if for every node v, there exist at least $k-1$ other nodes in the graph with the same degree as v [15]. They introduce practical anonymization algorithms that are based on principles related to the realizability of degree sequences.

Another approach was introduces by Backstrom, Dwork, and Kleinberg [2]. They consider several possible types of "injection" attacks, in which the intruder is actively involved in the social network before its data will be published in a repository, such that the intruder will be capable to retrieve his own data and to use it as a marker that facilitates the attack. Backstrom's work does not propose a practical method to counter the mentioned attacks.

6   Conclusions and Future Work

In this paper we studied a new anonymization approach for social network data. We introduced a generalization method for edges and a measure to quantify structural information loss. We developed a greedy privacy algorithm that anonymizes a social network. This algorithm can be user-balanced towards preserving more the structural information of the network or the nodes' attribute values.

We envision several research directions that can extend this work:

- Extend the anonymity model to achieve protection against attribute disclosure in social networks. Similar models such as p-sensitive k-anonymity [26], l-diversity [17], (α, k)-anonymity [30], and t-closeness [14] exist for microdata.
- Study the change in utility of an anonymized social network for various application fields.
- Formally analyze how the similarity measure is tied to the total structural information loss measure and improve the greedy selection criteria.

References

1. Bader, D.A., Madduri, K.: GTGraph: A Synthetic Graph Generator Suite (2006), http://www.cc.gatech.edu/~kamesh/GTgraph/
2. Backstrom, L., Dwork, C., Kleinberg, J.: Wherefore Art Thou R3579X? Anonymized Social Networks, Hidden Patterns, and Structural Steganography. In: International World Wide Web Conference (WWW), pp. 181–190 (2007)
3. Bamba, B., Liu, L., Pesti, P., Wang, T.: Supporting Anonymous Location Queries in Mobile Environments with PrivacyGrid. In: ACM World Wide Web Conference (2008)
4. Byun, J.W., Kamra, A., Bertino, E., Li, N.: Efficient k-Anonymization using Clustering Techniques. In: Kotagiri, R., Radha Krishna, P., Mohania, M., Nantajeewarawat, E. (eds.) DASFAA 2007. LNCS, vol. 4443, pp. 188–200. Springer, Heidelberg (2007)
5. Chakrabarti, D., Zhan, Y., Faloutsos, C.: R-MAT: A Recursive Model for Graph Mining. In: SIAM International Conference on Data Mining (2004)
6. Ciriani, V., Vimercati, S.C., Foresti, S., Samarati, P.: K-Anonymity. In: Secure Data Management In Decentralized Systems, pp. 323–353 (2007)
7. Ghinita, G., Karras, P., Kalinis, P., Mamoulis, N.: Fast Data Anonymization with Low Information Loss. In: Very Large Data Base Conference (VLDB), pp. 758–769 (2007)
8. Han, J., Kamber, M.: Data Mining. In: Concepts and Techniques, 2nd edn. Morgan Kaufmann, San Francisco (2006)
9. Hay, M., Miklau, G., Jensen, D., Weiss, P., Srivastava, S.: Anonymizing Social Networks. Technical Report No. 07-19, University of Massachusetts Amherst (2007)
10. Hay, M., Miklau, G., Jensen, D., Towsley, D., Weis, P.: Resisting Structural Re-identification in Anonymized Social Networks. In: Very Large Data Base Conference (VLDB), pp. 102–114 (2008)
11. HIPAA. Health Insurance Portability and Accountability Act (2002), http://www.hhs.gov/ocr/hipaa
12. Lambert, D.: Measures of Disclosure Risk and Harm. Journal of Official Statistics 9, 313–331 (1993)
13. LeFevre, K., DeWitt, D., Ramakrishnan, R.: Mondrian Multidimensional K-Anonymity. In: IEEE International Conference of Data Engineering (ICDE), vol. 25 (2006)
14. Li, N., Li, T., Venkatasubramanian, S.: T-Closeness: Privacy Beyond k-Anonymity and l-Diversity. In: IEEE International Conference on Data Engineering (ICDE), pp. 106–115 (2007)

15. Liu, K., Terzi, E.: Towards Identity Anonymization on Graphs. In: ACM SIGMOD International Conference on Management of Data, pp. 93–106 (2008)
16. Lunacek, M., Whitley, D., Ray, I.: A Crossover Operator for the k-Anonymity Problem. In: Genetic and Evolutionary Computation Conference (GECCO), pp. 1713–1720 (2006)
17. Machanavajjhala, A., Gehrke, J., Kifer, D.: L-Diversity: Privacy beyond K-Anonymity. In: IEEE International Conference on Data Engineering (ICDE), vol. 24 (2006)
18. Malin, B.: An Evaluation of the Current State of Genomic Data Privacy Protection Technology and a Roadmap for the Future. Journal of the American Medical Informatics Association 12(1), 28–34 (2005)
19. Meyerson, A., Williams, R.: On the complexity of optimal k-anonymity. In: ACM PODS Symposium on the Principles of Database Systems, pp. 223–228 (2004)
20. Newman, D.J., Hettich, S., Blake, C.L., Merz, C.J.: UCI Repository of Machine Learning Databases (1998),
    http://www.ics.uci.edu/~mlearn/MLRepository.html
21. Potterat, J.J., Philips-Plummer, L., Muth, S.Q., Rothenberg, R.B., Woodhouse, D.E., Maldonado-Long, T.S., Zimmerman, H.P., Muth, J.B.: Risk Network Structure in the Early Epidemic Phase of HIV Transmission in Colorado Springs. Sexually Transmitted Infections 78, 159–163 (2002)
22. Samarati, P.: Protecting Respondents Identities in Microdata Release. IEEE Transactions on Knowledge and Data Engineering 13(6), 1010–1027 (2001)
23. Shetty, J., Adibi, J.: The Enron Email Dataset Database Schema and Brief Statistical Report (2004),
    http://www.isi.edu/~adibi/Enron/Enron_Dataset_Report.pdf
24. Sweeney, L.: K-Anonymity: A Model for Protecting Privacy. International Journal on Uncertainty, Fuzziness, and Knowledge-based Systems 10(5), 557–570 (2002)
25. Sweeney, L.: Achieving k-Anonymity Privacy Protection Using Generalization and Suppression. International Journal on Uncertainty, Fuzziness, and Knowledge-based Systems 10(5), 571–588 (2002)
26. Truta, T.M., Bindu, V.: Privacy Protection: P-Sensitive K-Anonymity Property. In: PDM Workshop, with IEEE International Conference on Data Engineering (ICDE), vol. 94 (2006)
27. Tse, H.: An Ethnography of Social Networks in Cyberspace: The Facebook Phenomenon. The Hong Kong Anthropologist 2, 53–77 (2008)
28. Ward, H.: Prevention Strategies for Sexually Transmitted Infections: Importance of Sexual Network Structure and Epidemic Phase. Sexually Transmitted Infections 83, 43–49 (2007)
29. Wang, T., Liu, L.: Butterfly: Protecting Output Privacy in Stream Mining. In: IEEE International Conference on Data Engineering (ICDE), pp. 1170–1179 (2008)
30. Wong, R.C.W., Li, J., Fu, A.W.C., Wang, K.: (α, k)-Anonymity: An Enhanced k-Anonymity Model for Privacy-Preserving Data Publishing. In: SIGKDD, pp. 754–759 (2006)
31. Zheleva, E., Getoor, L.: Preserving the Privacy of Sensitive Relationships in Graph Data. In: ACM SIGKDD Workshop on Privacy, Security, and Trust in KDD (PinKDD), pp. 153–171 (2007)
32. Zhou, B., Pei, J.: Preserving Privacy in Social Networks against Neighborhood Attacks. In: IEEE International Conference on Data Engineering (ICDE), pp. 506–515 (2008)

Composing Miners to Develop an Intrusion Detection Solution

Marcello Castellano, Giuseppe Mastronardi,
Luca Pisciotta, and Gianfranco Tarricone

Dipartimento di Elettrotecnica ed Elettronica
Politecnico di Bari
Via Orabona 4, 70125 Bari, Italy
castellano@poliba.it

Abstract. Today, security is of strategic importance for many computer science applications. Unfortunately, an optimal solution does not exist and often system administrators are faced with new security problems when trying to protect computing resources within a reasonable time. Security applications that seem effective at first, could actually be unsuitable. This paper introduces a way of developing flexible computer security solutions which can allow system administrators to intervene rapidly on systems by adapting not only existing solutions but new ones as well. To this end, the study suggests considering the problem of intrusion detection as a Knowledge Discovery process and to describe it in terms of both e-services and miner building blocks. In addition, a definition of an intrusion detection process using Web content analysis generated by users is presented.

Keywords: Knowledge Discovery, Intrusion Detection, Mining Engine, Data mining, Web mining.

1 Introduction

In security, it is axiomatic that what cannot be prevented, must be detected. The goal of Intrusion Detection is to detect security violations in computer systems. Intrusion Detection is a passive approach to security as it monitors information systems and sounds alarms when security violations are detected. Examples of violations include the abuse of privileges or the use of attacks to exploit software or protocol vulnerabilities [1,2,3,4].

The aim of an Intrusion Detection System IDS is to support a security manager or a security management system in making decisions and signalling discovery of any possible intrusion. Various approaches for recognizing a possible intrusion have been presented in the literature. The use of interval timers and event counters to build time-series models have been proposed to build statistical models used for comparing short-term behaviour with long-term historical norms based on temporal regularities in audit data. Web usage patterns analysis from Web log files can also allow a better understanding of user behaviour on an individual or group basis by forming user profiles. In

F. Bonchi et al. (Eds.): PinkDD 2008, LNCS 5456, pp. 55–73, 2009.

most of the recent Intrusion Detection Systems, security administrators are able to extend the detection ability of the system by writing new rules which are usually hand-coded. This same process can also be carried out automatically by an intelligent IDS. Audit data based approaches are largely used to build intrusion detection models. Data must be labelled as either "normal" or "attack" in order to define the suitable behavioural models that represent these two different categories. By starting from a collection of user profiles, IDSs work to extract patterns, in order to create new knowledge for security managers [5,6,7,8,9,10,11].

Although signaling and identifying techniques are essential tools for an IDS, the most strategic element for building real efficiency in security lies in the administrators and analysts ability to rapidly adapt observation conditions and quickly establish the conditions for signaling any new anomalies which may occur in the system. These IDS user requirements define how an IDS must operate and thus influence system-architecture studies. Various authors have proposed specific solutions and frameworks for building efficient IDS. Many have suggested using Service Oriented Architecture (SOA) to develop security architecture. The main idea behind SOA is to provide services rather than access to systems. Trust management based solutions have been proposed to develop a security framework to build reasonable trust relationships among network entities. Moreover, to provide a collaborative environment to discover, verify and validate new IT system security requirements structured goal-oriented agent-based process modelling frameworks have been proposed. In addition, adaptable security solutions for large-scale systems based on security specification language in order to specify dynamic security policies have been proposed. Building IDSs by composable replaceable security services have been taken into account, also. [12,13,14,15,16,17]. This work presents a new architectural approach based on a mining engine which enables a rapid formation for intrusion detection solutions based on e-services for IDS.[18,19,20]

Knowledge Discovery is a complex and interdisciplinary field which deals with the understanding of unsuspected patterns and rules in data. Discovering these relationships is a highly interactive process where, in order to achieve positive results, the user must be able to apply various techniques or some given parameters on a permanent basis [21,22,23]. In order to create Knowledge Discovery Process different approaches have been proposed up to now. Today, the most effective strategy uses Mining Techniques. Data Mining searches for patterns in data which are interesting, according to user-defined measures of interestingness, and valid, according to user defined measures of validity. This area of research has recently gained much attention by industries. This is due to the existence of large collections of data in different formats, including large data warehouses, and the increasing need for data analysis and understanding. In addition to the mining of data stored in data warehouses, there has recently been increased interest in Web text mining, a process which can discover useful information from the content of web pages [24,25,26,27].

This paper will describe a process of Knowledge Discovery that can be managed flexibly by security administrators in order to quickly create new intrusion detection approaches. To obtain this result, Knowledge Discovery is proposed in terms of miners which are reusable, elementary components designed to be orchestrated for intrusion detection. We begin by introducing a Knowledge Discovery process model, and then describe our reference architecture for Knowledge Discovery using a mining approach. In the fourth section, we discuss the development of an Intrusion Detection solution in terms of miners. Finally, we present an example of the solution to define an intrusion detection process based on web-user content analysis using the proposed Intrusion Detection architecture.

2 Knowledge Discovery Process

A reference model of the Knowledge Discovery process based on e-services for Knowledge can be traced back to the Knowledge Discovery in Databases (KDD) model proposed by Piatesky-Shapiro, Matheus and Chan. The KDD has been defined as a nontrivial process for identifying valid, novel, potentially useful, and ultimately understandable patterns and, thus, is useful for extracting information from large collections of data. This process allows the identification of hidden structures within data by combining fields of databases and data warehousing with algorithms from machine learning and methods from statistics [28,29].

A generic method of Knowledge Discovery consists of the following tasks:

- Data Selection: data relevant to the analysis are retrieved from the database.

- Data Pre-Processing: noise and inconsistent data are removed (data cleaning) and multiple data sources are combined (data integration).

- Data Transformation: data are transformed or consolidated into forms appropriate for mining summary or aggregation operations.

- Data Mining: intelligent methods are applied in order to extract data patterns.

- Pattern Evaluation: interesting patterns representing knowledge based on measures of interest are identified.

- Knowledge Presentation: mined knowledge is presented to the user through visualization and knowledge representation techniques.

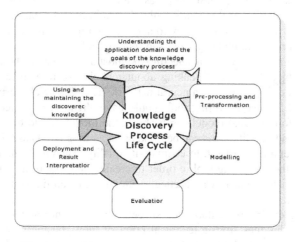

Fig. 1. Life Cycle of the Knowledge Discovery Process

The first step of a Knowledge Discovery process requires understanding a customer is domain objectives. First, the details to be collected and all problem areas (Marketing, Customer Care, Business Development, etc.) should be identified. In this step an inventory of resources, as well as requirements, assumptions, constraints, risks, contingencies, terminology, costs and benefits can be determined. In addition the business objective can be qualified in technical terms, adding details about Data Mining and Web Mining goals.

The next step involves project planning which includes the duration of the project, necessary resources, inputs, outputs and dependencies. Moreover, it is necessary at this point identify the collection of the data listed in the project resources in order to understand the application domain.

After this data set should be created by selecting and then integrating the collected data. This data set will represent the necessary input for the following data preparation steps. The collected and selected data should then be described, (their format, quantity and any other features discovered), and explored (e.g. by analyzing the distribution of key attributes, the relations between numbers of attributes or simple statistical analyses). This could improve Data and Web Mining processes and refine the data description and quality reports needed for further analysis, considering problem such as completeness, correctness, missing value or representation.

The pre-processing and transformation phase requires a preliminary study of data quality and technical constraints. This will provide a better understanding and selection of data for the analysis, thus, producing a final data set which reflects the precision and requirements requested. A good quality can be achieved by examining the data selected in order to extract clean subsets, with the possibility of inserting suitable default values or estimating missing data through other techniques. By preparing and integrating cleaned data sets, derived attributes, new records/values, and summarization, starting from multiple tables or records that have different information about the same object can be created.

Finally, it may be necessary to format transformed data without changing their meaning. Frequently, dataset records are initially organized in a given order but then the following modelling algorithm requires them to be arranged in a quite different order. The main outcome of this phase is the description of the dataset. This description also includes the actions necessary to address data quality issues and the process by which it was produced, table by table and field by field.

The modelling phase consists of selecting the most appropriate technique to achieve the business's objective. When speaking about models, we refer to a specific algorithm, such as a decisional tree or a neural network, as well as a clustering or an association method. Moreover, a fitting tool should have already been selected. After the selection and before the real model can be built, the model should be tested to determine its quality and validity. This can be done by separating the dataset into a train set, to build the model and a test set, to estimate its quality. Therefore, the model will be built by running the modelling tool in order to create one or more mining models and to support the decision makers with particular attention to their domain knowledge and the desired results.

While mining engineers evaluate the successfulness of the modelling application and discovery techniques from a technical point of view, business analysts and domain experts later discuss knowledge results in a business context. The output of the

Modelling step should describe how models are built, tested and evaluated and the process by which they are produced. The final outcome of model testing can be combined into one report that includes the Modelling assumption, Test design, Model description and Model assessment.

In the Evaluation phase, the degree to which the models extracted fit the business objective is estimated. If the result is poor, the reasons why the model has not fit the objectives properly and what may have been overlooked can be evaluated. An alternative evaluation method is to test the model on real applications and analyze the results. After reviewing the process, the decision to move on to deployment, begin further iterations, or set up new mining projects should be made.

The summary presented states whether a project meets its initial business objectives and provides hints and actions for activities that may have been missed and/or should be repeated. The relevant features are:

- Assessment of data mining results, aimed at comparing results with business objectives and criteria;

- Process review, aimed at establishing a possible repetition of the project;

- List of possible actions, aimed at detailing the next steps in the project.

The next phase consists of deployment and result interpretation. This could consist either in a simple summary of the project and its tests or in a final and comprehensive presentation of the mining results. In any case, the important results obtained are:

- Deployment plan, aimed at describing data and web mining results;

- Monitoring and maintenance plan, aimed at specifying how the results deployed are to be maintained;

- Final report, aimed at summarizing the project's global results, such as process, evaluation, different plans, cost/benefit analysis and conclusions.

In order to use and maintain the knowledge discovered, the project requires a detailed process monitoring plan. In this phase, a careful preparation of a maintenance strategy can help to avoid long periods of incorrect usage of mining results.

3   Reference Architecture for Knowledge Discovery Using a Mining Approach

Thus, a Mining Engine can be used as a reference architecture for Knowledge Discovery [18,19]. The purpose of a mining engine is to generate e-services and make then available for knowledge by driving the user through the main stages of the Knowledge Discovery process. Figure 2 shows an UML scheme that describes logic functions of the Mining Engine architecture and how components of the Mining Engine work together in this architecture.

The boxes on the left represent logic levels and explain the relation among the components while the boxes on the right describe the logic functions of each level. In detail, the e-Knowledge services represent a business level where decisions are taken. In this level, all the services that the Mining Engine has to provide are present. Then

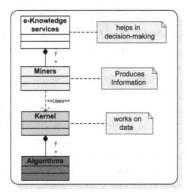

Fig. 2. Logic functions of the Mining Engine architecture

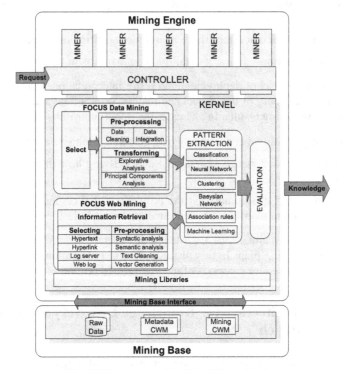

Fig. 3. The logic model of the Mining Engine

there are the Miners which are situated at the level where information is produced. They describe the operations that produce a service. The Kernel is the core of the system. It covers the process of Knowledge Discovery by working on structured and unstructured data and building new mining models. Finally, one or several data and web mining algorithms can be used to provide the kernel with the techniques to operate on the data.

Figure 3 shows the logic model of the Mining Engine. It consists of a collection of components, each one of which with a well defined task. This mining engine has been designed in order to work in a distributed data and web mining environment. An important component of the Mining Engine is the Controller. The Controller receives the request for the e-service for knowledge and then activates the business logic of the correspondent service by calling one or more Miners to provide the result. Miners are building blocks that can be used to build a complex application. In this case, the application is represented by an e-service knowledge.

Miners are situated at the level where information is produced and represent the operations that produce a service. They can either work by loading from the Mining Base the mining models associated to the required e-service for knowledge or by activating a training process of KDD or KDT in the Kernel according to the typology of the service or the mining model to be created. The Kernel follows the process of knowledge discovering starting from raw data and involving iterations of the main stages: data preparation, data and web mining, and results analysis. Finally, the Mining Base represents the knowledge repository of the whole architecture and its functions are those of repository for raw data, knowledge metadata and mining model.

3.1 Modeling Miner

Nowadays, business processes require well-defined building block components in order to create large structured applications. Service Oriented Architecture (SOA) represents a valid solution to meet these needs. The purpose of this architecture is to address the requirements of loosely-coupled, standard-based, and protocol-independent distributed computing by mapping enterprise information systems isomorphically with respect the overall business process flow. The SOA allows the composition of a set of independent but cooperating subsystems or services in order to create an application. The SOA model isolates each service and exposes only those interfaces which can be useful to other services. Consequently, as technology changes, services can be updated independently, thus limiting the impact of changes and updates to a manageable level.

Today, e-services for knowledge have to be quickly adapted to a decision maker's needs and be able to cover the full lifecycle of the Knowledge Discovery process. The creation of e-services for knowledge requires supporting the interactions among the different components of a mining architecture and the assembling of their logic into a manageable application. To overcome this difficulty, the SOA provides a flexible architecture that modularizes the Knowledge Discovery process into Miners. For "Miners" we mean replaceable and reusable components in a Knowledge Service Oriented Architecture (K-SOA) with the following characteristics:

- All operations in the Knowledge Discovery process are defined as Miners.
- All Miners are autonomous. The implementation and execution of the operations required are encapsulated behind the Miner Interface.

Here the Knowledge Discovery process is defined by applying the Cross Industry Standard Process for Data Mining methodology (CRISP-DM) [30,31,32]. CRISP-DM is described in terms of a hierarchical model consisting of four levels of abstraction, from general to specific: phase, generic task, specialized task and process instance

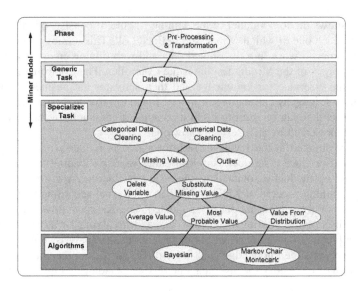

Fig. 4. An example of Miner Model for the phase of Pre-Processing & Transformation

(Figure 4). At the first level, the Knowledge Discovery process is organized into a number of phases; each phase then consists of several generic tasks. The second level is meant to be general enough to cover all possible Data and Web Mining situations and applications. Each Miner has to be as complete and stable as possible, and this particularly to make the model valid for yet unforeseen developments such as new modelling techniques or upgrading. The third level describes how to use Miners in certain specific situations.

An example of a miner model for the phase of pre-processing and transformation is shown in the Figure 4. Here, it is clear how a generic data cleaning task at the second level could be implemented in different situations, such as the cleaning of categorical values or numeric values.

3.2 Miner Orchestration

Knowledge allows the definition of reusable mining components (Miners) and their orchestration in order to provide more complex functionalities in the Knowledge Discovery process. Orchestration describes how services can interact with each other at the message level, including the business logic and execution order of the interactions. These interactions may span applications and/or organizations, and result in a long-lived, transactional, multi-step process model. Orchestration logic can be as simple as a single two-way conversation or as complex as a nonlinear multi-step transaction. It allows both a synchronous request-reply programming model and a conversational model based on asynchronous interactions across loosely-coupled e-services. Orchestration is necessary for the building of a Knowledge Service Oriented Architecture. It also defines how to arrange specialized data and web mining tasks (Miners) in order to provide more complex processes.

We use the term Miner Orchestration to define the process able to create a workflow of building blocks for the production of e-services for Knowledge. Miner Orchestration must be dynamic, flexible and adaptable so as to meet the changing needs of an organization. Flexibility can be achieved by providing a clear separation between the process logic and the Miners employed. To achieve this separation we propose an orchestration engine. The engine handles the overall process workflow, calling the appropriate Miners and determining the next steps to be completed. An Orchestration Engine is able to manage, at the level of specialized task, a workflow of Miners in order to provide e-services for knowledge. The workflow follows the CRISP-DM steps and includes the understanding of the application domain and the usage and maintenance of the discovered knowledge.

A fundamental aspect of Miner Orchestration is reusability. For example in fact if a new e-service for knowledge were needed, it would be sufficient to insert a new Miner or to use Miners already implemented.

4 Developing an Intrusion Detection Application

The primary assumptions of intrusion detection are that the user and program activities are observable. for example, via system auditing mechanisms and more importantly, normal and intrusion activities have distinct behaviour patterns. Intrusion detection therefore includes these essential elements:

- Resources to be protected in a target system, for example, network services, user accounts, system kernels, etc.
- Models that characterize the normal or legitimate behaviour of the activities involving these resources.
- Techniques that compare the observed activities with the established models. The activities that are not normal are flagged as intrusive.

Up to now, two different approaches for data mining based intrusion detection techniques have been proposed: misuse detection and anomaly detection [33,34].

In misuse detection, each instance in a data set is labelled as 'normal' or 'intrusion' and a learning algorithm is trained over the labelled data. These techniques allow automatic retraining of Intrusion Detection models on different input data, including new types of attacks, as long as they have been labelled appropriately. Unlike signature-based intrusion detection systems, models of misuse are created automatically, and can be more sophisticated and precise than manually created signatures. The misuse of detection techniques are characterized by their high degree of accuracy in detecting known attacks and their variations. Those techniques allow the detection attacks whose instances have not yet been observed.

Anomaly detection, on the other hand, builds models of normal behaviour, and automatically detects any deviation from it, flagging the latter as suspect. Anomaly detection techniques thus identify new types of intrusions as deviations from normal usage. A potential draw-back of these techniques is the rate of false alarms. This can happen primarily because previously unseen, yet legitimate, system behaviours may also be recognized as anomalies, and hence flagged as potential intrusions [11,22]. Precisely, the main problem related to both anomaly and misuse detection techniques resides in the encoded models, which define normal or malicious behaviours.

A pre-processing phase on raw audit data is necessary to fill records with a set of features, e.g., duration, source and destination hosts and ports, number of bytes transmitted, etc. Data mining algorithms are then applied to compute the frequent activity patterns, in the forms of association rules and frequent episodes, from the audit records. Association rules describe correlations among system features such as what shell command is associated with what argument. Instead, frequent episodes capture the sequential temporal co-occurrences of system events such as what network connections are made within a short time-span. Together, association rules and frequent episodes form the statistical summaries of system activities. When sufficient normal audit data is gathered, a set of frequent patterns can be computed and maintained as baseline.

Several types of algorithms are particularly useful for mining audit data:

• Classification places data items into one of several predefined categories. The algorithms normally output classifiers, for example, in the form of decision trees. In Intrusion Detection, one would want to gather sufficient normal and abnormal data and then apply a classification algorithm to teach a classifier to label or predict new, unseen audit data as belonging to the normal or abnormal class.

• Link analysis determines relations between fields in the database records.

• Correlation of system features in audit data, for example, between command and argument in the shell command history data of a user, can serve as the basis for constructing normal usage profiles.

• Sequence analysis models sequential patterns.

These algorithms can determine what time-based sequence of audit events occur frequently together. These frequent event patterns provide guidelines for incorporating temporal statistical measures into intrusion detection models. For example, patterns from audit data containing network-based denial of service attacks suggest that several per host and per service measures should be included.

4.1 Intrusion Detection Components

This section illustrates how an Intrusion Detection application can be simply designed in the SOA architecture by creating an Orchestration of already implemented Miners as building blocks.

According to the CRISP-DM, the process of creation of Miner Orchestration can be divided into steps: Data Pre-processing and Transformation; Modelling; Evaluation and Deployment of Result. Each step is made up of one or more Miners collaborating with one another. The Data Pre-processing and Transformation step gives an understanding as to which data are necessary for implementation.

Two main approaches can be used to identify a useful data set for implementation. The first relies on simulating a real-world network scenario, the second builds the set using actual traffic. The first approach is usually adopted when applying pattern recognition techniques to intrusion detection. The most well-known dataset is the KDD Cup 1999 Data. This set was created in order to evaluate the ability of data mining algorithms to build predictive models able to distinguish between a normal behaviour and a abnormal one. The KDD Cup 1999 Data contains a set of "connection" records coming out from the elaboration of raw data. Each connection is labelled as either "normal" or "attack". The connection records are built from a set of higher-level connection features that are able to discern normal network activities from those which are illegal. The most important aspect is being able to effectively reproduce the behaviour of network traffic sources. Collecting real traffic can be considered a viable alternative approach for the construction of a traffic data set. Although it can prove effective in real-time intrusion detection, it still presents some problems. In particular, collecting the data set by means of real traffic requires a data pre-classification process. In fact, as stated before the pattern recognition process must have a data set in which packets are labelled as either "normal" or "attack". Indeed, no information is available in the real traffic to distinguish the normal activities from the malicious ones in order to label the data set [7,8,35,36].

After collecting the data, they must be prepared, cleaned, and stored in a data warehouse for the pre-processing and transformation. The latter consists of creating and cleaning the dataset that will represent the input for the following analysis.

The Modelling step implements data mining techniques to create a mining model of the Intrusion Detection service. A number of data mining techniques have been found useful in the context of misuse and anomaly detection. Among the most popular techniques are pattern recognition, association rules, clustering, support vector machines, decision tree, classification, and frequent episode programs.

The final phase of this step is developing a framework for applying these data mining techniques to build intrusion detection models. The framework consists of one of these techniques as well as a support environment that enables system builders to interactively and iteratively drive the process of constructing and evaluating detection models. The final product should be concise and intuitive and be able to apply classification rules that can detect intrusions.

Finally, the importance of the Evaluation and Deployment of Result step lies in assessing the application generated rules and deciding statistically acceptable, trivial, spurious, or just not relevant. One way to validate the rules discovered is to use validation operators that allow a human expert to validate large numbers of rules at a time with relatively little input from the expert. Validation operators can be Similarity-based rule grouping, Template-based rule filtering, or Redundant-rule elimination.

Our study has shown that Data Mining can be used to support and partially automate the investigation in an Intrusion Detection process. Moreover not all data mining techniques are suitable for heterogeneous environments. To overcome these limitations, we propose a SOA approach, thus, retraining each dataset with the best possible techniques in order to find all possible intrusions. On the other hand, this approach is more expensive because it requires a great number of mining interactions. Thus, when designing and configuring such a system, a preliminary phase of trade off evaluation would be necessary.

4.2 Intrusion Detection Miners Orchestration

The previous paragraph introduced Miners and Orchestration as a theoretical approach for the realization of an IDS. In our experiments, these elements were built using standards as Java, Web Services and BPEL4WS [37,38]. Figure 5 shows the BPEL model which represents the orchestration of different Miners that collaborate to accomplish the process of an Intrusion Detection service. Moreover, how the workflow

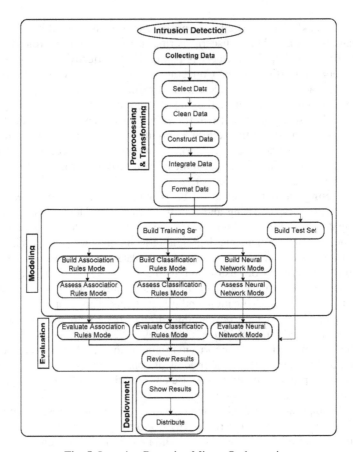

Fig. 5. Intrusion Detection Miners Orchestration

was organized in the main phases of Selecting, Data Pre-processing and Transformation, Modelling, Evaluation and Deployment of Result is illustrated. An initial set of data were retrieved, selected and then pre-processed in order to build the Training Set necessary for the following mine Modelling. Built models were tested and evaluated to assess their value and reliability and finally presented to the decision maker who analyzes the effectiveness of the knowledge extracted.

The following rows present the essential XML code of the relative BPEL process:

```
<?xml version="1.0" encoding="UTF-8"?>
<process expressionLanguage="Java"
name="IntrusionDetectionProcess"
<sequence name="MainSequence">
  <receive name="CollectingData">...</receive>
  <flow name="PreProcessingTrasformation">...
    <invoke name="SelectData">...</invoke>
    <invoke name="ConstructData">...</invoke>
    <invoke name="FormatData">...</invoke>
    <invoke name="CleanData">...</invoke>
```

```
        <invoke name="IntegrateData">...</invoke>
    </flow>
    <flow name="Modelling">...
        <invoke   name="BuildTrainingSet">...</invoke>
            <flow name="Build&Assess">...
                <invoke name="BuildAssociationRulesModel">...
                </invoke>
                <invoke name="BuildClassificationRulesModel">..
                </invoke>
                <invoke name="BuildNeuralNetworkModel">...
                </invoke>
                <invoke name="AssessAssociationRulesModel">...
                </invoke>
                <invoke name="AssessClassificationRulesModel">.
                </invoke>
                <invoke name="AssessNeuralNetworkModel">...
                </invoke>
            </flow>
        <invoke name="BuildTestSet">...</invoke>
    </flow>
    <flow name="Evaluation">...
        <invoke name="EvaluateAssociationRulesModel">...
        </invoke>
        <invoke name="EvaluateClassificationRulesModel">…
        </invoke>
        <invoke name="EvaluateNeuralNetworkModel">...
        </invoke>
        <invoke name="ReviewResults">...
        </invoke>
    </flow>
    <flow name="DeploymentofResults">...
        <invoke name="ShowResults">...</invoke>
        <reply name="Distribute">...</reply>
    </flow>
</sequence>
</process>
```

The sample Intrusion Detection process was created through miner orchestration. Java was used to produce the collection of Miner services and Weka technology was employed to support data mining techniques. The Controller was implemented with a BPEL4WS Engine and the Actions with BPEL4WS process definition files [39]. More details about these operations are available in our related works [18,19,20].

5 Experimental Results

The architectural model described here was used to produce a system prototype which could detect any possible intrusion and report it to the security administrator. Detection was carried out through the analysis of internet traffic generated by users in a network computer system. An intruder can be defined as a user able to overcome login phase illegitimately, and then surf the internet in an unauthorized manner using

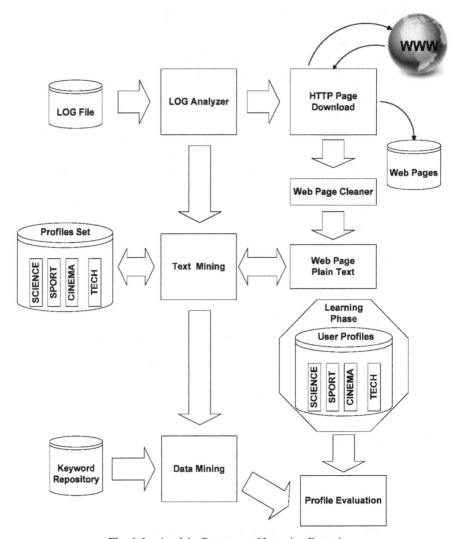

Fig. 6. Logic of the Prototype of Intrusion Detection

access credentials which are not his own. Since a user will normally generate web traffic based on his own habits and preferences a user profile can be created with data traffic analysis. The presence of an intruder can then be detected by comparing the traffic analysis with the user profile previously created.

The prototype created for this intrusion detection system is based on Knowledge Discovery techniques. It consists of various parts which work together to pursue a proposed goal. The system implementation is divided into several phases. In the first phase, useful data is collected to create a user model which can be analyzed. In this phase, a log file of the user's internet browsing habits is created. Next, a user profile is produced on the basis log file analysis. At this point, the user model is ready to be used to detect intrusions. If by analyzing the user log an intrusion is detected, a new user model should be created from this analysis. This more recent model must be similar to the original model. If it is not compatible within a specific tolerance with the original profile, then there is suspicion of an intrusion. Figure 6 shows a conceptual scheme of the system. The prototype was developed in JAVA. First, the system is launched with the aim of creating a user model to be analyzed and a browsing log which is typical of the user is created. The initial analysis then starts. Next, the web pages listed in the log are analyzed. Each page is automatically downloaded from the web module "http downloader". This module was created using the libraries JA-KARTA Apache. Then, the module "web page cleaner" extracts only the useful text content for analysis by text mining techniques and, from the inside, removes all scripts present. This will reproduce the page in plain text. The page is then sent to the module that deals with text mining. The "Text Mining" module has profiling rules that are based on a repository of thematic keywords.

On the plain text, previously obtained, a set of rules are applied in order to categorize the page in the log. The text mining system is realized using the GATE 4.0 toolkit [40]. The figure will show, for example, the text mining rule of the "sport" category. Then, at the end of the text mining phase, the number of topic keyword occurrences are saved. With this procedure, information that identifies the topics of the analyzed page are stored.

```
Phase: sport
Input: Token Lookup
Options: control = appelt
Rule: sport
({Lookup.majorType == sport}):sport
-->
{
gate.AnnotationSet sport=
    (gate.AnnotationSetBindings.get("sport");
gate.Annotation sportAnn =
    (gate.Annotation)sport.iterator().next();
gate.FeatureMap features =
    Factory.newFeatureMap();
features.put("rule", "sport");
outputAS.add(sport.firstNode(),sport.lastNode(),"Sport",

            features);
}
```

Fig. 7. GATE rule to identify the sport pages

These operations are repeated on every page in the user log, and the result is shown in the following figure.

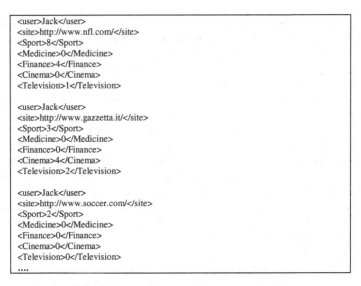

```
<user>Jack</user>
<site>http://www.nfl.com/</site>
<Sport>8</Sport>
<Medicine>0</Medicine>
<Finance>4</Finance>
<Cinema>0</Cinema>
<Television>1</Television>

<user>Jack</user>
<site>http://www.gazzetta.it/</site>
<Sport>3</Sport>
<Medicine>0</Medicine>
<Finance>0</Finance>
<Cinema>4</Cinema>
<Television>2</Television>

<user>Jack</user>
<site>http://www.soccer.com/</site>
<Sport>2</Sport>
<Medicine>0</Medicine>
<Finance>0</Finance>
<Cinema>0</Cinema>
<Television>0</Television>
....
```

Fig. 8. Repository gained from log analysis

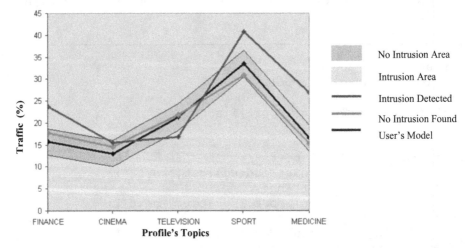

Fig. 9. Results from log analysis, in case of authorized navigation and unauthorized navigation

This procedure completely analyzes the user log. The modelling phase is completed by an analysis of the results using data mining techniques. This final stage receives the output of the previous phase as an input, and produces a result which is a profile of the user. In this final profile, each topic is associated with the percentage of occurrence inside the log files analyzed. Hence, the user profile consists of a series of topics which are associated with a percentage. This is the measure of a user's habit to visit pages about precise topics. At this point, the user profile is saved. To check a possible intrusion, the system runs in the previous scheme. In this case, however, the process starts from a new log file. This log file, is presumed to be generated by the user who had been previously analyzed, or in the case of intrusion by an intruder. This new analysis will generate a new user model/profile which is then compared to the original model. If the two models differ more than the tolerance defined by the system administrator, a suspected intrusion will be identified. To ensure that the system works well in time, we must consider the fact that a user's habits can change. Thus, a user's model must be periodically updated to guarantee its validity. Figure 9 shows a graphical representation of a user's profile generated by the system. The profile is represented by a broken line. This line connects the percentages associated with profile topics. The broken line is surrounded by a tolerance band. The tolerance level, represented by this band width, is defined by the system administrator. An intrusion can be recognized graphically. An intrusion occurs when a profile obtained from a subsequent analysis differs more than the fixed tolerance level allows. In the chart, the line indicating the intruder's user profile exits from the tolerance band.

6   Conclusions

This paper has discussed how to develop an intrusion detection application of a flexible security architecture based on e-services and miners. The architecture allows system administrators to intervene rapidly on new security requirement to study and develop more accurate security solutions. An intrusion detection approach using web content analysis strategy was presented.

References

1. Kemmerer, R.A., Vigna, G.: Intrusion Detection: A Brief History and Overview. Part supplement IEEE Computer 35(4), 27–30 (2002)
2. Tront, J.G., Marchany, R.C.: Internet security: intrusion detection & prevention. In: 37th IEEE Annual Hawaii International Conference on System Sciences, January 5-8. IEEE Press, New York (2004)
3. Smith, C.L., Robinson, M.: The Understanding of Security Technology and It's Application. In: IEEE 33rd Annual 1999 International Carnahan Conference on Security Technology, pp. 26–37. IEEE Press, New York (1999)
4. Kemmerer, R., Vigna, G.: Hi-DRA: Intrusion Detection for Internet Security. Proceedings of IEEE 93(10), 1848–1857 (2005)
5. Anderson, D., Lunt, T.G., Javitz, H., Tamaru, A., Valdes, A.: Detecting Unusual Program Behavior using the StatisticalComponents of the Next-Generation Intrusion Detection ExpertSystem (NIDES). In: Compuler Science Laboratory SRI-CSL-95-06 (May 1995)
6. Cai, Y., Clutter, D., Pape, G., Han, J., Welge, M., Auvil, L.: MAIDS: Mining Alarming Incidents from Data Streams. In: ACM-SIGMOD Int. Conf. Management of Data (SIGMOD 2004), pp. 919–920. ACM Press, New York (2004)
7. Mahoney, M.: A Machine Learning Approach to Detecting Attacks by Identifying Anomalies in Network Traffic. Florida Institute of Technology, Melbourne (2003)
8. Lee, W., Stolfo, S.J., Mok, K.W.: Mining Audit Data to Build ID Model. In: 4th International Conference on Knowledge Discovery and Data Mining, New York, pp. 66–72 (1998)
9. Seleznyov, A., Mazhelis, O., Puuronen, S.: Learning Temporal Regularities of User Behavior for Anomaly Detection. In: Gorodetski, V.I., Skormin, V.A., Popyack, L.J. (eds.) MMM-ACNS 2001. LNCS, vol. 2052, pp. 143–152. Springer, Heidelberg (2001)

10. Pepyne, D.L., Hu, J., Gong, W.: User Profiling for Computer Security. In: American Conference on Control, Boston, June 30 – July 2, pp. 982–987 (2004)

11. Esposito, M., Mazzariello, C., Oliviero, F., Romano, S.P., Sansone, C.: Real Time Detection of Novel Attacks by Means of Data Mining. In: ACM ICEIS Conference (2005)

12. Liu, Z., Campbell, R.H., Mickunas, M.D.: Security as services in active networks. In: Seventh International Symposium on Computers and Communications, pp. 883–890 (2002)

13. Torrellas, G.A.S., Cruz, D.V.: Security in a PKI-based networking environment: a multi-agent architecture for distributed security management system & control. In: Second IEEE International Conference on Computational Cybernatics, pp. 183–188 (2004)

14. Yau, S.S., Yao, Y., Chen, Z., Zhu, L.: An Adaptable Security Framework for Service-based Systems. In: 10th IEEE International Workshop on Object Oriented Real-Time Dependable Systems, pp. 28–35 (2005)

15. Yao, Z., Kim, D., Lee, I., Kim, K., Jang, J.: A security framework with trust management for sensor networks. In: Workshop of the 1st International Conference on Security and Privacy for Emerging Areas in Communication Networks, pp. 190–198 (2005)

16. Feiertag, R., Redmond, T., Rho, S.: A Framework for Building Composable Replaceable Security Services. In: DARPA Information Survivability Conference and Exposition. DISCEX 2000, vol. 2, pp. 391–402 (2000)

17. Chatzigiannakis, V., Androulidakis, G., Maglaris, B.: A Distributed Intrusion Detection Prototype Using Security Agents. In: 11th Workshop HP OpenView University Association (HPOVUA), Paris, France (June 2004)

18. Castellano, M., Pastore, N., Arcieri, F., Summo, V., Bellone de Grecis, G.: A Flexible Mining Architecture for Providing New E-Knowledge Services. In: 38th Annual Hawaii International Conference On System Sciences - Track 3. IEEE Computer Society Press, Los Alamitos (2005)

19. Castellano, M., Pastore, N., Arcieri, F., Summo, V., Bellone de Grecis, G.: Orchestrating the Knowledge Discovery Process. In: E-Service Intelligence: Methodologies, Technologies and Application. Springer, Berlin (2007)

20. Castellano, M., Mastronardi, G., Aprile, A., Minardi, M., Catalano, P., Dicensi, V., Tarricone, G.: A Decision Support System base line Flexible Architecture to Intrusion Detection. Journal of Software 2(6), 30–41 (2007)

21. Matheus, C.J., Chan, P.K., Piatetsky-Shapiro, G.: System for Knowledge Discovery in Databases. IEEE Transactions on Knowledge and Data Engineering (TKDE), Special Issue on Learning & Discovery in Knowledge-Based Databases 5(6), 903–913 (1993)

22. Lee, W., Stolfo, S.J.: Combining Knowledge Discovery and Knowledge Engineering to Build IDSs. In: 2nd International Workshop on Recent Advances in Intrusion Detection, West Lafayette, IN (1999)

23. WASET: 4th International Conference on Knowledge Mining. In: Proceedings of World Academy of Science, Engineering and Technology, vol. 26 (2007)

24. Fayyad, U.M., Piatetsky-Shapiro, G., Smith, P., Uthurusamy, R.: Advances in Knowledge Discovery and Data mining. MIT Press, London (1996)

25. Han, J., Kamber, M.: Data Mining: Concepts and Technique. Morgan Kaufmann Publishers, Academic Press, USA (2001)

26. Cooley, R., Mobasher, B., Srivastava, J.: Web Mining: Information and Pattern Discovery on the World Wide Web. In: Ninth IEEE International Conference on Tools with Artificial Intelligence, pp. 558–567. IEEE Press, New York (1997)

27. Zhang, W., Tang, X.: Web Text Mining on XSSC. In: Gu, J.F., Nakamori, Y., Wang, Z.T., Tang, X.J. (eds.) KSS 2006, pp. 167–175. Global Link Publisher (2006)

28. Felici, G., Vercellis, C.: Special Issue in Mathematical Method for Learning. Advances in Data Mining and Knowledge MML (2004); In: Computational optimization and Applications, vol. 38(2). Springer, Netherlands (2007)

29. Bozdogan, H.: Statistical Data Mining and Knowledge Discovery. Chapman and Hall/CRC, Boca Raton (2004)
30. CRoss Industry Standard Process for Data Mining, http://www.crisp-dm.org/
31. Chapman, P., Clinton, J., Kerber, R., Khabaza, T., Reinartz, T., Shearer, C., Wirth, R.: CRISP-DM 1.0 Step-by-step data mining guide. CRISP-DM Consortium. SPSS Inc. (2000), http://www.crisp-dm.org/CRISPWP-0800.pdf
32. Wirth, R., Hipp, J.: CRISP-DM: Towards a Standard Process Model for Data Mining. In: 4th International Conference on the Practical Applications of Knowledge Discovery and Data Mining (PADD 2000), Manchester, UK, pp. 29–39 (2000)
33. Lee, W., Stolfo, S.J., Mok, K.W.: Data mining approaches for intrusion detection. In: 7th USENIX Security Symposium, San Antonio, TX (1998)
34. Julisch, K.: Data mining for Intrusion Detection: a Critical Review. In: Barbara, D., Jajodia, S. (eds.) Applications of Data Mining in Computer Security. Kluwer Academic Publisher, Dordrecht (2002)
35. Paxson, V., Floyd, S.: Difficulties in simulating the internet. Transactions on Networking 9, 392–403 (2001)
36. Hackathom, R.D.: Web Farming for the Data Warehouse. In: Gray, J. (Series ed.) The Morgan Kaufmann Series in Data Management Systems (1998)
37. IBM, BEA Systems, Microsoft, SAP AG, Siebel Systems: Business Process Execution Language for Web Services (BPEL4WS), http://www.ibm.com/developerworks/library/specification/ws-bpel/
38. IBM, BEA Systems, Microsoft, SAP AG, Siebel: SystemsBusiness Process Execution Language for Web Services: Version 1.1, http://download.boulder.ibm.com/ibmdl/pub/software/dw/specs/ws-bpel/ws-bpel.pdf
39. Peltz, C.: Web Service Orchestration: a review of emerging technologies, tools, and standards.Techical report, Hewlett-Packard Company (2003)
40. GATE – General Architetcture for Text Engineering, http://gate.ac.uk/

Malicious Code Detection Using Active Learning

Robert Moskovitch, Nir Nissim, and Yuval Elovici

Deutsche Telekom Laboratories at Ben Gurion University
Ben Gurion University, Beer Sheva 84105, Israel
{robertmo,nirni,elovici}@bgu.ac.il

Abstract. The recent growth in network usage has motivated the creation of new malicious code for various purposes, including economic and other malicious purposes. Currently, dozens of new malicious codes are created every day and this number is expected to increase in the coming years. Today's signature-based anti-viruses and heuristic-based methods are accurate, but cannot detect new malicious code. Recently, classification algorithms were used successfully for the detection of malicious code. We present a complete methodology for the detection of unknown malicious code, inspired by text categorization concepts. However, this approach can be exploited further to achieve a more accurate and efficient acquisition method of unknown malicious files. We use an Active-Learning framework that enables the selection of the unknown files for fast acquisition. We performed an extensive evaluation of a test collection consisting of more than 30,000 files. We present a rigorous evaluation setup, consisting of real-life scenarios, in which the malicious file content is expected to be low, at about 10% of the files in the stream. We define specific evaluation measures based on the known precision and recall measures, which show the accuracy of the acquisition process and the improvement in the classifier resulting from the efficient acquisition process.

1   Introduction

The term malicious code (malcode) commonly refers to pieces of code, not necessarily executable files, which are intended to harm, generally or in particular, the specific owner of the host. Malcodes are classified, based mainly on their transport mechanism, into five main categories: worms, viruses, Trojans, and a new group that is becoming more common, which comprises remote access Trojans and backdoors.

The recent growth in high-speed internet connections and internet network services has led to an increase in the creation of new malicious codes for various purposes, based on economic, political, criminal or terrorist motives (among others). Some of these codes have been used to gather information, such as passwords and credit card numbers, as well as for behavior monitoring. A recent survey by McAfee[1] indicates that about 4% of search results from the major

---

[1] McAfee Study done by Frederick in june 2007
$http://www.newsfactor.com/story.xhtml?story_id = 010000CEUEQO$

F. Bonchi et al. (Eds.): PinkDD 2008, LNCS 5456, pp. 74–91, 2009.
© Springer-Verlag Berlin Heidelberg 2009

search engines on the web contain malicious code. Additionally, Shin et al. [1] found that above 15% of the files in the KaZaA network contained malicious code. Thus, we assume that the proportion of malicious files in real life is about or less than 10%, but we also consider other options of malicious files proportions through the imbalanced problem that will be briefly explain in the next section.

Current anti-virus technology is primarily based on two approaches. Signature-based methods, which rely on the identification of unique strings in the binary code, while being very precise, are useless against unknown malicious code. The second approach involves heuristic-based methods, which are based on rules defined by experts, which define a malicious behavior, or a benign behavior, in order to enable the detection of unknown malcodes [2]. Other proposed methods include behavior blockers, which attempt to detect sequences of events in operating systems, and integrity checkers, which periodically check for changes in files and disks. However, besides the fact that these methods can be bypassed by viruses, their main drawback is that, by definition, they can only detect the presence of a malcode after the infected program has been executed, unlike the signature-based methods, including the heuristic-based methods, which are very time-consuming and have a relatively high false alarm rate.

Recently, classification algorithms were employed to automate and extend the idea of heuristic-based methods. As we will describe in more detail shortly, the binary code of a file is represented by n-grams, and classifiers are applied to learn patterns in the code and classify large amounts of data. A classifier is a rule set which is learnt from a given training-set, including examples of classes, both malicious and benign files in our case. Recent studies, which we survey in the next section, have shown that this is a very successful strategy.

Another problem which is troubling the anti virus community is the acquisition of new malicious files, which is very important to detect as quickly as possible. This is often done by using honey-pots. Another option is to scan the traffic at the internet service provider, if accessible, to increase the probability of detection of a new malcode. However, the main challenge in both options is to scan all the files efficiently, especially when scanning internet node (router) traffic.

We present a methodology for malcode categorization based on concepts from text categorization. We present an extensive and rigorous evaluation of many factors in the methodology, based on SVM classifiers using three types of kernels. The evaluation is based on a test collection containing more than 30,000 files.

In this study we focus on the problem of efficiently scanning and acquiring new malicious code in a stream of executable files using Active Learners. We have shown that by using an inteligent acquisition of executables it is possible to acquire only small part of the files in the stream and still achieve significant improvement in the detection performance, an improvement that contributes to the learner's updatability in light of the new files over the net. while also saves time and money as will be explained later.

We start with a survey of previous relevant studies. We describe the methods we used to represent the executable files. We present our approach of acquiring new malcodes using Active Learning and perform a rigorous evaluation. Finally, we present our results and discuss them.

2   Background

2.1   Detecting Malcodes via Data Mining

Over the past five years, several studies have investigated the option of detecting unknown malcode based on its binary code. Schultz et al. [3] were the first to introduce the idea of applying machine learning (ML) methods for the detection of different malcodes based on their respective binary codes. They used three different feature extraction (FE) approaches – program header, string features, and byte sequence features – in which they applied four classifiers – a signature-based method (anti-virus), Ripper, a rule-based learner, Naive Bayes, and Multi-Naive Bayes. This study found that all the ML methods were more accurate than the signature-based algorithm. The ML methods were more than twice as accurate, with the out-performing method being Naive Bayes, using strings, or Multi-Naive Bayes using byte sequences.

Abou-Assaleh et al. [4] introduced a framework that used the common n-gram (CNG) method and the k nearest neighbor (KNN) classifier for the detection of malcodes. For each class, malicious and benign, a representative profile was constructed and assigned a new executable file. This executable file was compared with the profiles and matched to the most similar. Two different datasets were used: the I-worm collection, which consisted of 292 Windows internet worms, and the win32 collection, which consisted of 493 Windows viruses. The best results were achieved using 3-6 n-grams and a profile of 500-5000 features.

Kolter and Maloof [5] presented a collection that included 1971 benign and 1651 malicious executables files. N-grams were extracted and 500 were selected using the information gain measure [6]. The vector of n-gram features was binary, presenting the presence or absence of a feature in the file and ignoring the frequency of feature appearances. In their experiment, they trained several classifiers: IBK (KNN), a similarity based classifier called TFIDF classifier, Naive Bayes, SVM (SMO), and Decision tree (J48), the last three of which were also boosted. Two main experiments were conducted on two different datasets, a small collection and a large collection. The small collection consisted of 476 malicious and 561 benign executables and the larger collection of 1651 malicious and 1971 benign executables. In both experiments, the four best-performing classifiers were Boosted J48, SVM, boosted SVM, and IBK. Boosted J48 out-performed the others, The authors indicated that the results of their n-gram study were better than those presented by Schultz and Eskin [3].

Recently, Kolter and Maloof [7] reported an extension of their work, in which they classified malcodes into families (classes) based on the functions in their respective payloads. In the categorization task of multiple classifications, the best results were achieved for the classes: mass mailer, backdoor, and virus (no benign classes). In attempts to estimate their ability to detect malicious codes based on their issue dates, these classifiers were trained on files issued before July 2003, and then tested on 291 files issued from that point in time through August 2004. The results were, as expected, not as good as those of previous experiments. These results indicate the importance of maintaining such a training set through the acquisition of new executables, in order to cope with unknown new executables.

Henchiri and Japkowicz [8] presented a hierarchical feature selection approach which makes possible the selection of n-gram features that appear at rates above a specified threshold in a specific virus family, as well as in more than a minimal amount of virus classes (families). They applied several classifiers, ID3, J48 Naive Bayes, SVM- and SMO, to the dataset used by Schultz et al. [3] and obtained results that were better than those obtained using a traditional feature selection, as presented in [3], which focused mainly on 5-grams. However, it is not clear whether these results are reflective more of the feature selection method or of the number of features that were used.

Moskovitch et al [9] presented a test collection consisting of more than 30,000 executable files, which is the largest known to us. They performed a wide evaluation consisting of five types of classifiers and focused on the imbalance problem in real life conditions, in which the percentage of malicious files is less than 10%, based on recent surveys. After evaluating the classifiers on varying percentages of malicious files in the training set and test sets, it was shown to achieve the optimal results when having similar proportions in the training set as expected in the test set.

2.2 Active Learning and Selective Sampling

A major challenge in supervised learning is labeling the examples in the dataset. Often the labeling is expensive since it is done manually by human experts. Labeled examples are crucial in order to train a classifier, and we would therefore like to reduce the number of labeling requirements. The Active Learning (AL) approach proposes a method which asks actively for labeling of specific examples, based on their potential contribution to the learning process.

AL is roughly divided into two major approaches: the membership queries [10] and the selective-sampling approach [11]. In the membership queries approach the learner constructs artificial examples from the problem space, then asks for its label from the expert, and finally learns from it and so forth, in an attempt to cover the problem space and to have a minimal number of examples that represent most of the types among the existing examples. However, a potential practical problem in this approach is requesting a label for a nonsense example. The selective-sampling approach usually comprises a pool-based sampling, in which the learner is given a large set of unlabeled data (pool) from which it iteratively selects the most informative and contributive examples for labeling and learning, based on which it is carefully selects the next examples, until it meets stopping criteria.

Studies in several domains successfully applied active learning in order to reduce the effort of labeling examples. Unlike in random learning, in which a classifier is trained on a pool of labeled examples, the classifier actively indicates the specific examples that should be labeled, which are commonly the most informative examples for the training task. Two AL methods were considered in our experiments: Simple-Margin Tong and Koller [12] Error-Reduction Roy and McCallum [13].

2.3 Acquisition of New Malicious Code Using Active Learning

As we presented briefly earlier the option of acquiring new malicious files from the web and internet services providers is essential for fast detection and updating of the anti-viruses, as well as updating of the classifiers. However, manually inspecting each potentially malicious file is time-consuming, and often done by human experts. We propose using Active Learning as a selective sampling approach based on a static analysis of malicious code, in which the active learner identifies new examples which are expected to be unknown. Moreover, the active learner is expected to present a ranked list of the most informative examples, which are probably the most different from what currently is known.

3   Methods

3.1   Text Categorization

To detect and acquire unknown malicious code, we suggest implementing well-studied concepts from the information retrieval (IR) and more specific text categorization domain. In execution of our task, binary files (executables) are parsed and n-gram terms are extracted. Each n-gram term in our task is analogous to words in the textual domain. Here are descriptions of the IR concepts used in this study. Salton and Weng [14] presented the vector space model to represent a textual file as a bag-of-words. After parsing the text and extracting the words, a vocabulary of the entire collection of words is constructed. Each of these words may appear zero to multiple times in a document. A vector of terms is created, such that each index in the vector represents the term frequency (TF) in the document. Equation (1) shows the definition of a normalized TF, in which the term frequency is divided by the maximal appearing term in the document with values in the range of [0-1]. Another common representation is the TF Inverse Document Frequency (TFIDF), which combines the frequency of a term in the document (TF) and its frequency in the documents collection, as shown in Equation (2), in which the term's (normalized) TF value is multiplied by the $IDF = log(\frac{N}{n})$, where N is the number of documents in the entire file collection and n is the number of documents in which the term appears.

$$TF = \frac{term\ frequency}{max(term\ frequency\ in\ document)} \tag{1}$$

$$TFIDF = TF * log(DF), \quad Where\ DF = \frac{N}{n} \tag{2}$$

3.2   Data Set Creation

We created a dataset of malicious and benign executables for the Windows operating system, which is the most commonly used and attacked. To the best of our knowledge, this collection is the largest ever assembled. We acquired the malicious files from the VX Heaven website[2] , having 7688 malicious files. To identify the files, we used the Kaspersky anti-virus and the Windows version of the Unix 'file' command for file type identification. The files in the benign set, including executable and Dynamic Linked Library (DLL) files, were gathered from machines running the Windows XP operating system, which is currently considered the most used, on our campus. The benign set contained 22,735 files, which were reported by the Kaspersky anti-virus[3] program as being completely virus-free.

3.3   Data Preparation and Feature Selection

We parsed the binary code of the executable files using several n-gram lengths moving windows, denoted by n. Vocabularies of 16,777,216, 1,084,793,035, 1,575,804,954, and 1,936,342,220, for 3-gram, 4-gram, 5-gram and 6-gram, respectively, were extracted. Later the TF and TFIDF representation were calculated for each n-gram in each file.

In machine learning applications, the large number of features (many of which do not contribute to the accuracy and may even decrease it) in many domains presents a huge problem. Moreover, in our task a reduction in the amount of features is crucial for practical reasons, but must be performed while simultaneously maintaining a high level of accuracy. This is due to the fact that, as shown earlier, the vocabulary size may exceed billions of features, far more than can be processed by any feature selection tool within a reasonable period of time. Additionally, it is important to identify those terms that appear in most of the files, in order to avoid zeroed representation vectors. Thus, initially the features having the highest DF value Equation (2) were extracted.

Based on the DF measure, two sets were selected, the top 5,500 terms and the top 1,000-6,500 terms. The set of top 1000 to 6,500 set of features was inspired by the removal of stop-words, as often done in information retrieval for common words. Later, feature selection methods were applied to each of these two sets. Since it is not the focus of this paper, we will describe the feature selection preprocessing very briefly. We used a filters approach, in which the measure was independent of any classification algorithm, to compare the performances of the different classification algorithms. In a filters approach, a measure is used to quantify the correlation of each feature to the class (malicious or benign) and estimate its expected contribution to the classification task. Three feature selection measures were used: as a baseline we used the document frequency measure DF (Equation 2), and additionally the Gain Ratio (GR) [6] and Fisher Score [15]. Eventually the top 50, 100, 200 300, 1000, 1500 and 2000 were selected from each feature selection.

[2] http://vx.netlux.org
[3] http://www.kaspersky.com

3.4 Support Vector Machines

We employed the SVM classification algorithm using three different kernel functions, in a supervised learning approach. We briefly introduce the SVM classification algorithm and the principles and implementation of Active Learning that we used in this study. SVM is a binary classifier which finds a linear hyperplane that separates the given examples into the two given classes. Later an extension that enables handling multiclass classification was developed. SVM is widely known for its capacity to handle a large amount of features, such as text, as was shown by Joachims [16]. We used the Lib-SVM implementation of Chang [17] that also handles multiclass classification.

Given a training set, in which an example is a vector $x_i =< f_1, f_2, ...f_m >$, where f_i is a feature, and labeled by $y_i = -1, +1$, the SVM attempts to specify a linear hyperplane that has the maximal margin, defined by the maximal (perpendicular) distance between the examples of the two classes. Fig. 1 illustrates a two dimensional space, in which the examples are located according to their features and the hyperplane splits them according to their label.

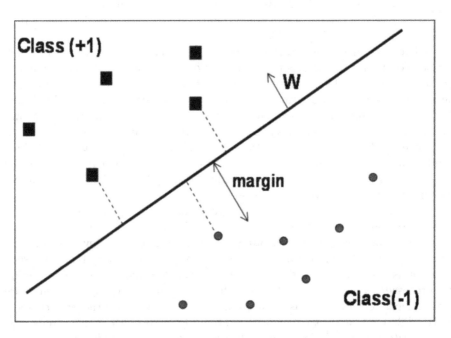

Fig. 1. An SVM that separates the training set into two classes, having maximal margin in a two dimensional space

The examples lying closest to the hyperplane are the "supporting vectors" W, the Normal of the hyperplane, is a linear combination of the most important examples (supporting vectors), multiplied by LaGrange multipliers (alphas). Since the dataset in the original space often cannot be linearly separated, a kernel function K is used. SVM actually projects the examples into a higher dimensional space in order to create linear separation of the examples. Note that when the kernel function satisfies Mercer's condition, as was explained by Burges [18], K can be written as shown in Equation (3), where ϕ is a function that maps the example from the original feature space into a higher dimensional space, while K relies on "inner product" between the mappings of examples x_1, x_2. For the general case, the SVM classifier will be in the form shown in Equation (4), while n is the number of examples in training set, and w is defined in Equation (5).

$$K(x_1, x_2) = \phi(x_1) \cdot \phi(x_2) \tag{3}$$

$$F(x) = sign(W \cdot \phi(x)) = sign(\sum_{1}^{n} \alpha_i y_i K(x_i, x)) \tag{4}$$

$$W = \sum_{1}^{n} \alpha_i y_i \phi(x_i) \tag{5}$$

Two commonly used kernel functions were used: Polynomial kernel, as shown in Equation (6), creates polynomial values of degree p, where the output depends on the direction of the two vectors, examples x_1, x_2, in the original problem space. Note that a private case of a polynomial kernel, having p=1, is actually the Linear kernel. Radial Basis Function (RBF), as shown in Equation (7), in which a Gaussian is used as the RBF and the output of the kernel depends on the Euclidean distance of examples x_1, x_2.

$$K(x_1, x_2) = (x_1 \cdot x_2)^p \tag{6}$$

$$K(x_1, x_2) = e(-\frac{|x_1 - x_2|^2}{2\sigma^2}) \tag{7}$$

3.5   Active Learning

In this study we implemented two selective sampling (pool-based) AL methods: the Simple Margin presented by Tong and Koller [12] ,and Error Reduction presented by Roy and McCallum [13].

Simple-Margin. This method is directly oriented to the SVM classifier. As was explained in the section 3.4, by using a kernel function, the SVM implicitly projects the training examples into a different (usually higher dimensional) feature space, denoted by F. In this space there is a set of hypotheses that are consistent with the training-set, meaning that they create linear separation of the training-set. This set of consistent hypotheses is called Version-Space (VS).

From among the consistent hypotheses, the SVM then identifies the best hypothesis that has the maximal margin. Thus, the motivation of the Simple-Margin AL method is to select those examples from the pool, so that these will reduce the number of hypotheses in the VS, in an attempt to achieve a situation where VS contains the most accurate and consistent hypotheses. Calculating the VS is complex and impractical when large datasets are considered, and therefore this method is oriented through simple heuristics that are based on the relation between the VS and the SVM with the maximal margin. Practically, examples that lie closest to the separating hyperplane (inside the margin) are more likely to be informative and new to the classifier, and thus will be selected for labeling and acquisition.

Error-Reduction. The Error Reduction method is more general and can be applied to any classifier that can provide probabilistic values for its classification decision. Based on the estimation of the expected error, which is achieved through adding an example into the training-set with each label, the example that is expected to lead to the minimal expected error will be selected and labeled. Since the real future error rates are unknown, the learner utilizes its current classifier in order to estimate those errors.

In the beginning of an AL trial, an initial classifier $\hat{P}_D(y|x)$ is trained over a randomly selected initial set D. For every optional label $y\epsilon Y$ (of every example x in the pool P) the algorithm induces a new classifier $\hat{P}_{D'}(y|x)$ trained on the extended training set $D' = D + (x, y)$ Thus (in our binary case, malicious and benign are the only optional labels) for every example x there are two classifiers, each one for each label. Then for each one of the example's classifiers the future expected generalization error is estimated using a log-loss function, shown in Equation 8. The log-loss function measures the error degree of the current classifier over all the examples in the pool, where this classifier represents the induced classifier as a result of selecting a specific example from the pool and adding it to the training set, having a specific label. Thus, for every example $x\epsilon P$ we actually have two future generalization errors (one for each optional label as was calculated in Equation (8)). Finally, an average is calculated for the two errors, which is called the self-estimated average error, based on Equation (9).

It can be understood that it is built of the weighted average so that the weight of each error of example x with label y is given by the prior probability of the initial classifier to classify correctly example x with label y. Finally, the example x with the lowest expected self-estimated error is chosen and added to the training set. In a nutshell, an example will be chosen from the pool only if it dramatically improves the confidence of the current classifier more than all the examples in the pool (means lower estimated error).

$$E_{\hat{P}_{D'}} = \frac{1}{|P|} \sum_{x\epsilon X} \sum_{y\epsilon Y} \hat{P}_{D'}(y|x) \cdot |log(\hat{P}_{D'}(y|x))| \qquad (8)$$

$$\sum_{y\epsilon Y} \hat{P}_D(y|x) \cdot E_{\hat{P}_{D'}} \qquad (9)$$

4   Evaluation

To evaluate the use of AL in the task of efficient acquisition of new files, we defined specific measures derived from the experimental objectives. The first experimental objective was to determine the optimal settings of the term representation (TF or TFIDF), n-grams representation (3, 4, 5 or 6), the best global range (top 5500 or top 1000-6500) and feature selection method (DF, FS or GR), and the top selection (50, 100, 200, 300, 1000, 1500 or 2000). After determining the optimal settings, we performed a second experiment to evaluate our proposed acquisition process using the two AL methods.

4.1   Evaluation Measures

For evaluation purposes, we measured the True Positive Rate (TPR) measure, which is the number of positive instances classified correctly, as shown in Equation (10), False Positive Rate (FPR), which is the number of negative instances misclassified Equation (10), and the Total Accuracy, which measures the number of absolutely correctly classified instances, either positive or negative, divided by the entire number of instances, shown in Equation (11).

$$TPR = \frac{|TP|}{|TP + FN|} \quad , FPR = \frac{|FP|}{|FP + TN|} \tag{10}$$

$$Total\_Accuracy = \frac{|TP + TN|}{|TP + FP + TN + FN|} \tag{11}$$

4.2   Evaluation Measures for the Acquisition Process

In this study we wanted to evaluate the acquisition performance of the Active-Learner from a stream of files presented by the test set, containing benign and malicious executables, including new (unknown) and not-new files. Actually, the task here is to evaluate the capability of the module to acquire the new files in the test set, which cannot be evaluated only by the common measures evaluated earlier. Figure 2 illustrates the evaluation scheme describing the varying contents of the test set and Acquisition set that will be explained shortly. The datasets contain two types of files: Malicious (M) and Benign (B). While the Malicious region is presented as a bit smaller, it is actually significantly smaller. These datasets contain varying files partially known to the classifier, from the training set, and a larger portion of New (N) files, which are expected to be acquired by the Active Learner, illustrated by a circle. The active learner acquires from the stream part of the files, illustrated by the Acquired (A) circle. Ideally the Acquired circle will be identical to the New circle.

To define the evaluation measures, we define the resultant regions in the evaluation scheme by:

- $A \cap M \backslash N$ - The Malicious files Acquired, but not New.

- $A \cap M \cap N$ - The Malicious files Acquired and New.

- $M \cap N \backslash A$- The New Malicious files, but not Acquired.

- $A \cap B \backslash N$ - The Benign files Acquired, but not New.

- $A \cap B \cap N$ - The Benign files Acquired and New.

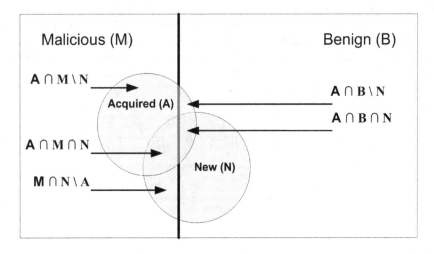

Fig. 2. An illustration of the evaluation scheme, including the Malicious (M) and Benign (B) Files, the New files to acquire (N) and the actual Acquired (A) files

For the evaluation of the said scheme we used the known Precision and Recall measures, often used in information retrieval and text categorization. We first define the traditional precision and recall measures. Equation (12) represents the Precision, which is the proportion of the accurately classified examples among the classified examples. Equation (13) represents the Recall measure, which is the proportion of the classified examples from a specific class in the entire class examples.

$$Precision = \frac{|\{Relevant\ examples\}| \cap |\{Classified\ examples\}|}{|\{Classified\ examples\}|} \qquad (12)$$

$$Recall = \frac{|\{Relevant\ examples\}| \cap |\{Classified\ examples\}|}{|\{Relevant\ examples\}|} \qquad (13)$$

As we will elaborate later, the acquisition evaluation set will contain both malicious and benign files, partially new (were not in the training set) and partially not-new (appeared in the training set), and thus unknown to the classifier. To evaluate the selective method we define here the precision and recall measures in the context of our problem. Corresponding to the evaluation scheme presented in Figure 2, the *precision_new_benign* is defined in Equation (14) by the proportion among the new benign files which were acquired and the acquired benign files. Similarly the *precision_new_malicious* is defined in Equation (15). The *recall_new_benign* is defined in Equation (16) by how many new benign files in the stream were acquired from the entire set of new benign in the stream. The *recall_new_malicious* is defined similarly in Equation (17).

$$Precision\_new\_benign = \frac{A \cap B \cap N}{A \cap B} \qquad (14)$$

$$Precision\_new\_malicious = \frac{A \cap M \cap N}{A \cap M} \qquad (15)$$

$$Recall\_new\_benign = \frac{A \cap B \cap N}{N \cap B} \qquad (16)$$

$$Recall\_new\_malicious = \frac{A \cap M \cap N}{N \cap M} \qquad (17)$$

The acquired examples are important for the incremental improvement of the classifier; The Active Learner acquires the new examples which are mostly important for the improvement of the classifier, but not all the new examples are acquired, especially these which the classifier is certain on their classification. However, we would like to be aware of any new files (especially malicious) in order to examine them and add them to the repository. This set of files are the New and not Acquired ($N \backslash A$), thus, we would like to measure the accuracy of the classification of these files to make sure that the classifier classified them correctly. This is done using the Accuracy measure as presented in Equation (11) on the subset defined by ($N \backslash A$), where for example $|TP(N \backslash A)|$ is the number of malicious executables that were labeled correctly as malicious, out of the unacquired new examples. In addition we measured the classification accuracy of the classifier in classifying examples which were not new and not acquired. Thus, using again the Accuracy measure (Equation (11)) for the $\neg(N \cup A)$ defines our evaluation measure.

5   Experiments and Results

5.1   Experiment 1

To determine the best settings of the file representation and the feature selection we performed a wide and comprehensive set of evaluation runs, including all the combinations of the optional settings for each of the aspects, amounting to

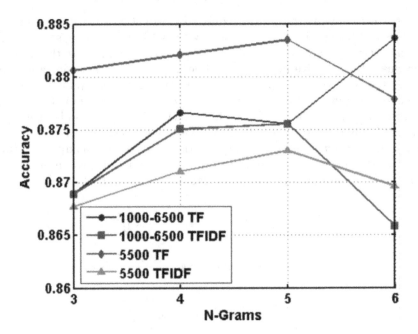

Fig. 3. The results of the global selection, term representation, and n-grams, in which the Top 5500 global selection having the TF representation is outperforming, especially with 5-grams

1536 runs in a 5-fold cross validation format for all the three kernels. Note that the files in the test-set were not in the training-set presenting unknown files to the classifier.

Global Feature Selection vs n-grams. Figure 3 presents the mean accuracy of the combinations of the term representations and n-grams. The top 5,500 features outperformed with the TF representation and the 5-gram in general. The out-performing of the TF has meaningful computational advantages, on which we will elaborate in the Discussion. In general, mostly the 5-grams outperformed the others.

Feature Selections and Top Selections. Figure 4 presents the mean accuracy of the three feature selection methods and the seven top selections. For fewer features, the FS outperforms, while above the Top 300 there was not much difference. However, in general the FS outperformed the other methods. For all the three feature selection methods there is a decrease in the accuracy when using above Top 1000 features.

Classifiers. After determining the best configuration of 5-Grams, Global top 5500, TF representation, Fischer score, and Top300, we present in Table 1 the results of each SVM kernel. The RBF kernel out-performed the others and had a low false positive rate, while the other kernels also perform very well.

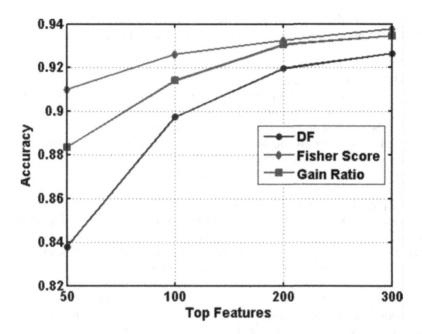

Fig. 4. The accuracy increased as more features were used, while in general the FS outperformed the other measures

Table 1. The RBF kernel outperformed while maintaining a low level of false positive

Kernel	Accuracy	FP	FN
SVM-LIN	0.921	0.033	0.214
SVM-POL	0.852	0.014	0.544
SVM-RBF	0.939	0.029	0.154

5.2   Experiment 2. Files Acquisition

In the second experiment we used the optimal settings from experiment 1, applying only the RBF kernel which outperformed (Table 1). In this set of experiments, we set an imbalanced representation of malicious-benign proportions in the test-set to reflect real life conditions of 10% malicious files in the stream, based on the information provided in the Introduction. In a previous study [9] we found that the optimal proportions in such scenario are similar settings in the training set. The Dataset includes 25000 executables (22,500 benign, 2500 malicious), having 10% malicious and: 90% benign contents as in real life conditions. The evaluation test collection included several components: Training-Set, Acquisition-Set (Stream), and Test-set. The Acquisition-set consisted of benign and malicious examples, including known executables (that appeared in the training set) and unknown executables (which did not appear in the training set) and the Test-set included the entire Data-set. These sets were used in the following steps of the experiment:

1. A Learner is trained on the Training-Set.
2. The model is tested on the Test-Set to measure the initial accuracy.
3. A stream of files is introduced to the Active Learner, which asks selectively for labeling of specific files, which are acquired.
4. After acquiring all the new informative examples, the Learner is trained on the new Training-Set.
5. The Learner is tested on the Test-Set.

We applied the learners in each step using 2 different variation of cross validation for each AL method. For the Simple-Margin we used variation of 10-fold cross validation. Thus, the Acquisition Set (stream) contained part of the folds in the Training Set and the Test Set, which was used for evaluation prior to the Acquisition phase and after, contained all the folds.

Simple-Margin AL method - We applied the Simple Margin Active Learner in the experimental setup presented earlier. Table 2 presents the mean results of the cross validation experiment. Both the Benign and the Malicious Precision were very high, above 99%, which means that most of the acquired files were indeed new. The Recall measures were quite low, especially the Benign Recall. This can be explained by the need of the Active Learner to improve the accuracy. An interesting fact is the difference in the Recall of the Malicious and the Benign, which can be explained by the varying proportions in the training set, which was 10% malicious. The classification accuracy of the new examples that were not acquired was very high as well, being close to 99%, which was also the classification accuracy of the not new, which was 100%. However, the improvement between the Initial and Final accuracy was significant, which shows the importance and the efficiency in the acquisition process.

Table 2. The Simple-Margin acquisition performance

Measure	Simple Margin Performance
Precision Benign	99.81%
Precision Malicious	99.22%
Recall Benign	33.63%
Recall Malicious	82.82%
$Accuracy(N\backslash A)$	98.90%
$Accuracy\neg(N \cap A)$	100%
Initial Accuracy on Test-Set	86.63%
Final Accuracy on Test-Set	92.13%
Number Examples in Stream	10250
Number of New Examples	7750
Number Examples Acquired	2931

Table 3. The Error-reduction acquisition performance

Measure	Error-reduction Performance
Precision Benign	97.563%
Precision Malicious	72.617%
Recall Benign	29.050%
Recall Malicious	75.676%
$Accuracy(N \backslash A)$	98.90%
$Accuracy \neg (N \cap A)$	100%
Initial Accuracy on Test-Set	85.803%
Final Accuracy on Test-Set	89.045%
Number Examples in Stream	3010
Number of New Examples	2016
Number Examples Acquired	761

Error-Reduction AL method - We performed the experiment using the Error Reduction method. Table 3 presents the mean results of the cross validation experiment. In the acquisition phase, the Benign Precision was high, while the Malicious Precision was relatively low, which means that almost 30% of the examples that were acquired were not actually new. The Recall measures were similar to those for the Simple-Margin, in which the Benign Recall was significantly lower than the Malicious Recall. The classification accuracy of the not acquired files was high both for the new and for the not new examples.

6   Discussion and Conclusion

We introduced the task of efficient acquisition of unknown malicious files in a stream of executable files. We proposed using Active Learning as a selective method for the acquisition of the most important files in the stream to improve the classifier's performance. This approach can be applied at a network communication node (router) at a network service provider to increase the probability of acquiring new malicious files. A methodology for the representation of malicious and benign executables for the task of unknown malicious code detection was presented, adopting ideas from Text Categorization.

In the first experiment, we found that the TFIDF representation has no added value over the TF, which is not the case in IR. This is very important, since using the TFIDF representation introduces some computational challenges in the maintenance of the measurements whenever the collection is updated. To reduce the number of n-gram features, which ranges from millions to billions, we used the DF threshold. We examined the concept of stop-words in IR in our domain and found that the top features have to be taken (e.g., top 5500 in our case), and not those of an intermediate level. Having the top features enables vectors which are less zeroed, since the selected features appear in most of the files. The Fisher Score feature selection outperformed the other methods, and using the top 300 features resulted in the best performance.

In the second experiment, we evaluated the proposed method of applying Active Learning for the acquisition of new malicious files. We examined two AL methods, Simple Margin and Error Reduction, and evaluated them rigorously using cross validation. The evaluation consisted of three main phases: training on the initial Training-set and testing on a Test-set, acquisition phase on a dataset including known files (which were presented in the training set) and new files, and eventually evaluating the classifier after the acquisition on the Test-set to demonstrate the improvement in the classifier performance.

For the acquisition phase evaluation we presented a set of measures based on the Precision and Recall measures dedicated for the said task, which refer to each portion of the dataset, the acquired benign and malicious, separately. For the not acquired files we evaluated the performance of the classifier in classifying them accurately to justify that indeed they did not need to be acquired. In general, both methods performed very well, with the Simple Margin performing better than the Error Reduction. In the acquisition phase, the benign and malicious Precision was often very high; however, the malicious Precision for the Error Reduction was relatively low. The benign and malicious Recalls were relatively low and reflected the classifier's needs.

An interesting phenomenon was that a significantly higher percentage of new malicious files, relatively to the benign files, were acquired. This can be explained by the imbalanced proportions of the malicious-benign files in the initial training set. The classification accuracy of the not acquired files, unknown and known, was extremely high in both experimental methods. The evaluation of the classifier before the acquisition (initial training set) and after showed an improvement in accuracy which justifies the process. However, the relatively low accuracy, unlike in the first experiment, can be explained by the small training set which resulted from the cross validation setup.

When applying such a method for practical purposes we propose that a human first examine the malicious acquired examples. However, note that there might be unknown files which were not acquired, since the classifier didn't consider them as informative enough and often had a good level of classification accuracy. However, these files should be acquired. In order to identify these files, one can apply an anti-virus on the files which were not acquired and were classified as malicious. The files which were not recognized by the anti-virus are suspected as unknown malicious files and should be examined and acquired.

Acknowledgments. We would like to thank Clint Feher, who created the dataset and Yuval Fledel for meaningful discussions and comments in the efficient implementation aspects.

References

1. Shin, S., Jung, J., Balakrishnan, H.: Malware Prevalence in the KaZaA File-Sharing Network. In: Internet Measurement Conference (IMC), Brazil (October 2006)
2. Gryaznov, D.: Scanners of the Year 2000: Heuristics. In: Proceedings of the 5th International Virus Bulletin (1999)

3. Schultz, M., Eskin, E., Zadok, E., Stolfo, S.: Data mining methods for detection of new malicious executables. In: Proceedings of the IEEE Symposium on Security and Privacy, pp. 178–184 (2001)
4. Abou-Assaleh, T., Cercone, N., Keselj, V., Sweidan, R.: N-gram Based Detection of New Malicious Code. In: Proceedings of the 28th Annual International Computer Software and Applications Conference, COMPSAC 2004 (2004)
5. Kolter, J.Z., Maloof, M.A.: Learning to detect malicious executables in the wild. In: Proceedings of the Tenth ACM SIGKDD International Conference on Knowledge Discovery and Data Mining, pp. 470–478. ACM, New York (2004)
6. Mitchell, T.: Machine Learning. McGraw-Hill, New York (1997)
7. Kolter, J., Maloof, M.: Learning to Detect and Classify Malicious Executables in the Wild. Journal of Machine Learning Research 7, 2721–2744 (2006)
8. Henchiri, O., Japkowicz, N.: A Feature Selection and Evaluation Scheme for Computer Virus Detection. In: Proceedings of ICDM 2006, pp. 891–895 (2006)
9. Moskovitch, R., Stopel, D., Feher, C., Nissim, N., Elovici, Y.: Unknown Malcode Detection via Text Categorization and the Imbalance Problem. In: IEEE Intelligence and Security Informatics (ISI 2008), Taiwan (2008)
10. Angluin, D.: Queries and concept learning. Machine Learning 2, 319–342 (1988)
11. Lewis, D., Gale, W.: A sequential algorithm for training text classifiers. In: Proceedingsof the Seventeenth Annual International ACM-SIGIR Conference on Research and Development in Information Retrieval, pp. 3–12. Springer, Heidelberg (1994)
12. Tong, S., Koller, D.: Support vector machine active learning with applications to text classification. Journal of Machine Learning Research 2, 45–66 (2000-2001)
13. Roy, N., McCallum, A.: Toward Optimal Active Learning through Sampling Estimation of Error Reduction. In: ICML (2001)
14. Salton, G., Wong, A., Yang, 'C.S.: A vector space model for automatic indexing. Communications of the ACM 18, 613–620 (1975)
15. Golub, T., Slonim, D., Tamaya, P., Huard, C., Gaasenbeek, M., Mesirov, J., Coller, H., Loh, M., Downing, J., Caligiuri, M., Bloomfield, C., Lander, E.: Molecular classification of cancer: Class discovery and class prediction by gene expression monitoring. Science 286, 531–537 (1999)
16. Joachims, T.: Making large-scale support vector machine learning practical. In: Scholkopf, B., Burges, C., Smola, A.J. (eds.) Advances in Kernel Methods: Support Vector Machines. MIT Press, Cambridge (1998)
17. Chang, C.C., Lin, C.-J.: LIBSVM: a library for support vector machines (2001), http://www.csie.ntu.edu.tw/~cjlin/libsvm
18. Burges, C.J.C.: A tutorial on support vector machines for pattern recognition. Data Mining and Knowledge Discovery 2, 121–167 (1988)

Maximizing Privacy under Data Distortion Constraints in Noise Perturbation Methods

Yaron Rachlin[1], Katharina Probst[2], and Rayid Ghani[1]

[1] Accenture Technology Labs, Chicago, IL, USA
yaron.rachlin@alumni.cmu.edu, rayid.ghani@accenture.com
[2] Google Inc., Atlanta, GA, USA
katharina.probst@gmail.com

Abstract. This paper introduces the 'guessing anonymity,' a definition of privacy for noise perturbation methods. This definition captures the difficulty of linking identity to a sanitized record using publicly available information. Importantly, this definition leads to analytical expressions that bound data privacy as a function of the noise perturbation parameters. Using these bounds, we can formulate optimization problems to describe the feasible tradeoffs between data distortion and privacy, without exhaustively searching the noise parameter space. This work addresses an important shortcoming of noise perturbation methods, by providing them with an intuitive definition of privacy analogous to the definition used in k-anonymity, and an analytical means for selecting parameters to achieve a desired level of privacy. At the same time, our work maintains the appealing aspects of noise perturbation methods, which have made them popular both in practice and as a subject of academic research.

Keywords: Noise perturbation, privacy, anonymity, statistical disclosure control.

1 Introduction

The tension between protecting privacy and preserving data utility is a fundamental problem for organizations that would like to share data. In cases where this problem is not resolved, data is either not shared, preventing useful applications, or organizations adopt risky practices of disclosing private information, sometimes with unfortunate results (as in the case of AOL [1]). The problem of modifying data to preserve data utility while protecting privacy is a broad area known as statistical disclosure control. This paper proposes a novel approach in the subdiscipline of statistical disclosure control that considers the anonymization of static individual data, known as microdata. Such data is commonly used in statistical analysis and data mining. The purpose of our approach is to enable data sharing while protecting the privacy of the individuals whose records comprise the shared data.

Domingo-Ferrer [2] and Aggarwal and Yu [3] survey a number of approaches for statistical disclosure control in microdata. Noise perturbation (also referred

F. Bonchi et al. (Eds.): PinkDD 2008, LNCS 5456, pp. 92–110, 2009.
© Springer-Verlag Berlin Heidelberg 2009

to as randomization) is a technique for anonymizing microdata. In this approach, data values are altered according to some probability distribution. Agrawal and Srikant [4], Agrawal and Aggarwal [5], and Muraldihar and Sarathy [6] investigate the privacy guarantees of randomization methods by considering the difficulty of estimating attribute values of individual records using sanitized data. While this represents one type of adversarial attack, work on the protection provided by noise-perturbation against an adversary that uses information in public databases to identify a sanitized record has been comparatively lacking. Samarati and Sweeney [7] introduced the k-anonymity model to directly address the danger of re-identifying a record using information in public databases. Methods that achieve k-anonymity reduce the granularity of the data, using operators such as generalization and suppression, so that every record is identical to at least $k - 1$ other records. Such methods rely on a combinatorial search, and do not use noise perturbation. Bayardo and Agrawal [8] and LeFevre et al. [9] introduce algorithms that achieve k-anonymity in such a manner.

This paper presents a definition of privacy for noise perturbation methods that addresses the problem of identification using publicly available information. We quantify privacy by formulating the identity linking attack as a guessing game. The privacy of a sanitized record, which we call the **guessing anonymity**, is defined by the number of guesses an attacker needs to correctly guess the original record used to generate the sanitized record. Recent work proposed conceptually similar but technically different definitions of privacy for noise perturbation methods [10,11,12]. These papers empirically studied the difficulty of linking for various noise models, but did not offer an approach to finding noise parameters that maximize the difficulty of linking anonymized records. In contrast, the definition of guessing anonymity enables us to explicitly relate the parameters of a noise perturbation model and the level of privacy using the the guessing inequality [13]. Using this definition, we formulate optimization problems that maximize privacy given constraints on the noise parameters. The constraints on noise parameters can be used to limit the degree to which the noise perturbation model will distort the data values. Recent work [14] proposed a probabilistic definition of k-anonymity suitable for quantifying privacy in noise perturbation methods. This paper proposed a method for achieving probabilistic k-anonymity for real attributes. We compare our approach to probabilistic k-anonymity in Section 4, and demonstrate that our approach can be used to find noise parameters that achieve a higher level of privacy with less noise. In contrast to [14], which provides a separate analysis for each noise model, our approach applies to general noise models and to both real and categorical attributes.

Before we proceed to describe our approach, we briefly outline several advantages of noise perturbation methods that render them particularly useful for data mining and analysis tasks. Our work preserves these advantages, while providing the additional advantage of protection from attacks based on public information. In noise perturbation, distributions are chosen in order to preserve data statistics, and noise models are often published along with the perturbed data. As a result, such methods are amenable to statistical analysis. For example, additive zero mean, uncorrelated Gaussian noise enables unbiased estimates of the mean of the data to which it is applied. Agrawal and Srikant [4] and Agrawal and Aggarwal [5] present methods for estimating the distribution of the original data using noise perturbed data. These papers demonstrate that noise-perturbation can be used to provide information about the statistics of the data, while limiting the disclosure of private information. In contrast, the statistical implications of approaches that do not use published noise modes, such as algorithms that achieve k-anonymity, are unclear.

An additional advantage of noise perturbation arises in many business applications. The research community is most familiar with organizations that release data for research purposes (e.g., AOL, Netflix, and KDD Cups). In these cases, the data set is used for both training and testing the models developed. A different situation arises when businesses outsource datamining tasks, such as building predictive models, to third-parties in order to develop solutions to specific business problems. Models built using anonymized shared data are incorporated back into the business processes of the client organization and applied to test data that is in an un-anonymized form. This becomes challenging when the shared training data was anonymized using techniques such as generalization-based k-anonymity algorithms that change the representation of the data. One motivation for the work presented in this paper was to develop anonymization techniques that will enable the training of models that can be applied to un-anonymized data without further modification.

We begin by defining the guessing anonymity and demonstrating the application of the guessing inequality in Section 2. This section presents the form of the guessing inequality for several useful noise perturbation models. Section 3 demonstrates how this inequality can be used to optimize the trade-off between anonymity and distortion, and how the tradeoff changes with the complexity of the noise perturbation model. Finally, in Section 4, we use our approach to anonymize the UCI Adult data set, and empirically evaluate the effect of anonymization on realistic data mining tasks such as classification and regression. We also compare our results to the approach described in [14].

Notation and Definitions: We denote random variables and functions by upper-case letters, and instantiations or constants by lower-case letters. Bold font denotes vectors. Sets are denoted using calligraphic script. The conditional Renyi entropy of a random variable X conditioned on Y is defined in [13] as

$$H_\alpha(X|Y) = \frac{\alpha}{1-\alpha} \ln \sum_y \left[\sum_x P_{XY}(x,y)^\alpha \right]^{\frac{1}{\alpha}} \tag{1}$$

for $\alpha > 0, \alpha \neq 1$. The Renyi entropy [15] is an information measure for which the Shannon entropy is a special case. One operational meaning for the conditional Renyi entropy is provided by the Guessing Inequality in Section 2.2.

2   Guessing Anonymity

How should one define privacy against attacks using publicly available information for noise perturbation methods? This is a key question that motivates this paper. We quantify privacy for noise perturbation methods by defining the guessing anonymity. Given a database of records and a sanitized version of this database, the guessing anonymity of a sanitized record is the number of guesses that an optimal attacker requires to link this sanitized record to a specific individual or entity, using publicly available data. In this section, we will define the guessing anonymity more precisely.

Table 1. Original Database

Name	Age	Procedure or Prescription
Alice	19	Antidepressants
Bob	15	Antibiotics
Chris	52	Chemotherapy
Diana	25	Abortion

Table 2. Sanitized Database

Name	Age	Procedure or Prescription	Guessing Anonymity
***	23.1	Antidepressants	2
***	19.4	Antibiotics	2
***	49.3	Chemotherapy	1
***	21.1	Abortion	2

We consider a structured database consisting of M records. To provide an example for our definitions, we use the fictitious health insurance claims database shown in Table 1. This databases contains four records, each with three attributes. We categorize the attributes as follows, using language introduced by [7,16]. The 'name' attribute, which would completely identify an individual, is an *identifier*. The 'age' attribute is a *quasi-identifier*, since such an attribute can be compared with external information sources to identify an individual in the database. The 'procedure or prescription' attribute is *confidential* because it contains sensitive information that is not assumed to be publicly available. Linking the identity and this confidential information (using the quasi-identifiers) is the goal of the attack we address in this paper. The quasi-identifiers associated with the record i are denoted r_i, with $i \in \mathcal{I} = \{1, \ldots, M\}$, where each i corresponds to some individual or entity. Each r_i is an N-dimensional vector, whose elements can be either real or categorical values. When sanitizing a record, the identifiers are suppressed and the sensitive attributes are preserved. The quasi-identifiers of record i, r_i, are sanitized and released as s (along with the unchanged sensitive information). We model privacy by quantifying the difficulty of linking

Table 3. Attacker Information

Name	Age	Procedure or Prescription
Alice	19	?
Chris	52	?

the quasi-identifiers of a sanitized record s with the true quasi-identifiers of the record from which it was generated using publicly available information. We assume that data sanitization of the quasi-identifiers is performed using noise perturbation, that is s is sampled from the distribution $P_{S|I}$. I is a random variable indicating record r_I, and the probability of drawing sanitized record s for record r_i is $P_{S|I}(S = s | I = i)$. The sanitized version of Table 1 is shown in Table 2. For each record, the age attribute in Table 1 is independently corrupted using additive zero mean Gaussian noise to obtain the age values in Table 2. In this example, $P_{S|I}$ is a Gaussian centered at the age associated with the record, and s is sampled from this distribution.

2.1   Definition

Given a released record s, we quantify the difficulty of linking it with a true record as follows. Define a guessing strategy as a sequence of questions of the form "Is r_i the vector of quasi-identifiers used to generate s?" Define the guessing function G as a function of the sanitized quasi-identifiers s, such that for each s, $G(\cdot, s) : \mathcal{I} \to \mathcal{I}$ is bijective and denotes the number of guesses required to guess r_i when an attacker observes s. We define the optimal guessing strategy, as the strategy that minimizes $E_{IS}[G(I|S)]$. As shown in [13], the optimal strategy guesses records in decreasing order of probability. That is, for any fixed sanitized record s, the optimal guessing strategy guesses the records in decreasing order based on their conditional probability $P(I = i | S = s)$ (i.e., the highest probability record is guessed first).

Definition. The **guessing anonymity** of the sanitized record s is the number of guesses that the optimal guessing strategy requires in order to correctly guess the record used to generate the sanitized record.

To illustrate this definition, consider an attacker with knowledge that Alice and Chris are in the database, and knowledge of their true age, as shown in Table 3. The guessing anonymity of each record is shown in the fourth column of Table 2. While the record that corresponds to Alice has a guessing anonymity of two, the sanitized record corresponding to Chris has a guessing anonymity of one due to the fact that his age is significantly higher than the ages in the other records. To protect the anonymity of Chris' record a Gaussian with a larger variance should have been used to sanitize his record. Note that the average guessing anonymity across these records is 1.75. Intuitively, the guessing anonymity of a record quantifies the difficulty of using the sanitized record to perform some attack. For example, consider an example where a record's confidential field contains a person's income, and an identity thief attempts to use this information to open a fake account. The guessing anonymity provides a metric of how many unsuccessful attempts the identity thief should expect to make before successfully opening the account. For a sufficiently high guessing anonymity, the number of attempts could set off an alarm.

We note that recent work introduced related definitions of privacy for noise perturbation methods. [10] introduced the definition of 'k-randomization.' According to this definition, a record is k-randomized if the number of invalid records that are a more likely match to the sanitized record than the original record is *at least* k. Though this notion is similar to our definition, our definition differs by not providing a lower limit on the number of invalid records that provide a more likely match, and by explicitly establishing a connection between privacy and guessing functions. These differences enable us to exploit the guessing inequality to establish a direct link between the noise parameters and the expected guessing anonymity in Section 2.2. [14] introduced the definition of probabilistic k-anonymity. A record is probabilistically k-anonymized if the *expected* number of invalid records that are a more likely match to the sanitized record than the true record is *at least* k. As in the case of k-randomization, this definition differs from the definition of guessing anonymity. [14] demonstrates an approach for achieving probabilistic k-anonymity, and we compare this approach to our work in Section 4.

2.2 Guessing Inequality

An important advantage of the definition of guessing anonymity is that it allows us to exploit an analytical relationship between perturbation noise and anonymity based on the application of the guessing inequality [13]. [17] first studied the relationship between entropy and the expected number of guesses. [13] introduced the guessing inequality to bound the moments of the guessing function. For $\rho > 0$, $\alpha = \frac{1}{1+\rho}$, we write the guessing inequality as follows,

$$H_\alpha(I|S) - \log(1 + \ln(M)) \leq \frac{1}{\rho} \log \left[\min_G E_{IS}[G(I|S)]^\rho \right] \leq H_\alpha(I|S) \quad (2)$$

The guessing inequality shows that the moments of the guessing function are upper and lower bounded by the conditional Renyi entropy. [13] points out that this result can be extended for the case where S takes values in a continuous space. In this case, (1) changes by replacing the sum over y with an integral.

We use (2) to bound the first moment (i.e. $\rho = 1$) of the guessing function for several different data and noise perturbation models. Assume a simple database consisting of M records, where each record is a single real number r_i (e.g., age in Table 1). Using the guessing inequality, we lower bound the expected number of guesses when we independently perturb each record with zero mean Gaussian noise of variance σ^2, i.e.,

$$P_{S|I}(s|i) = \frac{1}{\sqrt{2\pi\sigma^2}} e^{-\frac{(s-r_i)^2}{2\sigma^2}} \quad (3)$$

Since there are M records in the database $P_I = \frac{1}{M}$. Applying the guessing inequality to this simple model, and defining the constant $c \doteq \frac{1}{1+\ln(M)} \frac{1}{M}$, we have,

$$E\left[G(I|S)\right] \geq \frac{1}{(1+\ln(M))} \int_{-\infty}^{\infty} \left[\sum_{x=1}^{M} P_{SI}(s,i)^{\frac{1}{2}}\right]^2 ds$$

$$= c \sum_{i=1}^{M} \sum_{j=1}^{M} e^{-\frac{(r_i-r_j)^2}{8\sigma^2}} \doteq F(\sigma^2) \tag{4}$$

It is interesting to note that the lower bound on the expected number of guesses depends on the pairwise differences between all the records. This demonstrates that the bound relies on the actual values of the data being anonymized.

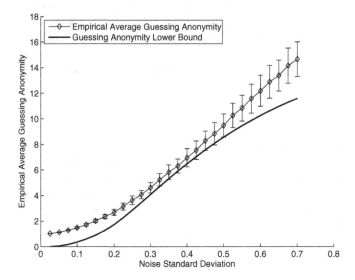

Fig. 1. Empirical guessing anonymity of database corrupted with Gaussian noise, as a function of noise level. The guessing inequality bounds the empirical average.

The guessing inequality provides an analytical connection between privacy and the parameters of the noise perturbation model. To demonstrate this connection, we consider the simple example of a database that consists of two hundred records, each with one real attribute. The records are generated by independently sampling a zero mean unit variance Gaussian distribution for each record. These records are then corrupted with Gaussian noise of a given variance. Figure 1 plots the average empirical guessing anonymity as a function of the standard deviation of the Gaussian used to corrupt the data. The empirical guessing anonymity of each record in the noise-corrupted data is measured by computing the number of guesses required to link the noise-corrupted record with the original noiseless record. We plot the average empirical guessing anonymity, where each point is obtained by generating two hundred noise-corrupted databases and averaging the empirical guessing anonymity across all records and databases. The error bars in the graph indicate an interval of two standard deviations. Figure 1 also contains a plot of the lower bound on the expected guessing anonymity shown in (4). Unlike the empirical curve, the lower bound does not require simulating a linking attack, and can be computed using only the parameters of our noise model.

We can also derive lower bounds for more complex noise and databases models. For example, assume that the database consists of M records each with N attributes, where each record r_i is a vector of real numbers. The expected number of guesses when we independently perturb each attribute k in each record with additive zero-mean Gaussian noise of variance σ_k^2 can be lower bounded as follows,

$$\mathrm{E}\left[G(I|S)\right] \geq c \sum_{i=1}^{M} \sum_{j=1}^{M} \prod_{k=1}^{N} e^{-\frac{(r_{ik}-r_{jk})^2}{8\sigma_k^2}} \tag{5}$$

This lower bound reduces to (4) when $N = 1$. We can provide a lower bound for a more complex noise model. We lower bound the expected number of guesses when we independently perturb each attribute k of the quasi-identifiers in record i with zero mean Gaussian noise of variance σ_{ik}^2.

$$\mathrm{E}\left[G(I|S)\right] \geq c \sum_{i=1}^{M} \sum_{j=1}^{M} \prod_{k=1}^{N} \sqrt{\frac{2\sigma_{ik}\sigma_{jk}}{\sigma_{ik}^2 + \sigma_{jk}^2}} e^{-\frac{(r_{ik}-r_{jk})^2}{4(\sigma_{ik}^2+\sigma_{jk}^2)}} \tag{6}$$

The equation above reduces to (5) when the variance is the same across records, i.e. $\sigma_{ik}^2 = \sigma_{jk}^2$ for all $i, j \in \mathcal{I}$. As can be seen in the sequence of lower bounds above, the guessing inequality can be computed for increasingly complex noise models. For example, a further extension could consider the effect of dependent perturbations. Given that independent Gaussian perturbation is widely used, we will not demonstrate additional lower bounds for other noise models for real data. Instead, we demonstrate that this lower bounds also applies to categorical variables.

In categorical data $r \in \mathcal{R}_1 \times \ldots \times \mathcal{R}_N$ and $s \in \mathcal{S}_1 \times \ldots \times \mathcal{S}_N$ where $\{\mathcal{R}_k, \mathcal{S}_k, k = 1, \ldots, N\}$ are finite sets. We assume a noise model where each quasi-identifier is perturbed independently, $P_{S|I}(s|i) = \prod_{k=1}^{N} P_{S_k|I}(s_k|r_{ik})$ and where $P_{S_k|I}(s_k|r_{ik}) = p_{sr}^k$ is a conditional probability distribution specifying transition probabilities from $\mathcal{R}_k$ to $\mathcal{S}_k$. For such a noise model,

$$\mathrm{E}\left[G(I|S)\right] \geq c \sum_{i=1}^{M} \sum_{j=1}^{M} \prod_{k=1}^{N} \sum_{h=1}^{|\mathcal{S}_k|} P_{S_k|I}(s_h|r_{ik})^{\frac{1}{2}} P_{S_k|I}(s_h|r_{jk})^{\frac{1}{2}} \tag{7}$$

We note that depending on the number of attributes and records, it may be more computationally efficient to compute (2) directly rather than the equation above. As for the case of Gaussian perturbation, we observe that (7) is a function of the values of the data set being sanitized. Finally, we note that a lower bound for mixed categorical and real attributes can be easily derived, for example, by combining (5) and (7).

3   Distortion and Privacy Tradeoff

The guessing inequality provides analytical bounds on the expected guessing anonymity of noise-perturbed data. In this section, we use this bound to optimize the trade-off between data utility and privacy. We define utility by considering the expected distortion between the original N attribute quasi-identifiers r and the noise-perturbed data s. Intuitively, high distortion means low utility, and vice versa. This definition of utility enables us to formulate and solve optimization problems to find the noise parameters that provide the maximum level of privacy given constraints on data distortion.

We define the distortion function generally as $d : \mathcal{R}_1 \times \ldots \times \mathcal{R}_N \times \mathcal{S}_1 \times \ldots \times \mathcal{S}_N \to \mathbb{R}^+$ In our experiments, we consider the following commonly used distortion functions. For **real-valued attributes**, i.e., $\mathcal{R}_k = \mathcal{S}_k = \mathbb{R}$ and $r, s \in \mathbb{R}^N$, we consider the mean squared error distortion,

$$d_{\mathrm{MSE}}(i, s) = \frac{1}{M} \sum_{i=1}^{M} \sum_{k=1}^{N} (r_{ik} - s_{ik})^2 \tag{8}$$

For **categorical data**, we consider the case when $\mathcal{R}_k$ is a finite set and $\mathcal{R}_k = \mathcal{S}_k$. The Hamming distortion is defined as follows,

$$d_{\mathrm{H}}(i, s) = \frac{1}{M} \sum_{i=1}^{M} \sum_{k=1}^{N} I(r_{ik}, s_{ik}) \tag{9}$$

where $I(r, s) = 1$ if $r \neq s$ and 0 otherwise. Naturally, one can define other distortion functions in an application-dependent manner. For example, distortion of some records or attributes can be weighted more heavily in situations where a domain expert assesses that certain attributes are more important than others.

Given the guessing inequality-based bounds on privacy and our definition of distortion, we can state an optimization problem whose solutions will enable us to show tradeoffs between distortion and privacy.

$$\max_{\substack{\theta \\ \mathrm{E}_{IS}[d(I,S)] \leq C}} (1 + \ln(M))^{-1} e^{H_{0.5}(I|S)} \tag{10}$$

The variables θ are the parameters of the noise perturbation model. The space of feasible noise parameters is defined by the constraint that the expected distortion must be less than C. We call this the **Distortion Constraint**. As a simple example, consider the form of (10) for the case of a database where each record consists of a single real quasi-identifier and a noise perturbation model where each record is independently perturbed using zero mean Gaussian noise with the same variance. We assume the distortion metric is the mean squared error shown in (8). (10) is now written as,

$$\max_{\substack{\sigma \\ \sigma^2 \leq C}} F(\sigma^2) \tag{11}$$

where $F(\sigma^2)$ is defined in (4). One could naturally study the related optimization problem, where the guessing inequality is used as a constraint, and one seeks to minimize distortion. We focus on this particular formulation due to the simplicity of the distortion-based constraints.

Work on noise perturbation methods such as [6] focuses on privacy from the perspective of the difficulty of estimating attributes using the sanitized data. To limit this risk, such research proposes constraints on the noise parameters. For example, in the case of Gaussian noise, that the variance must be greater than some constant. Though we did not pursue this direction of work in this paper, it would be fairly straightforward to incorporate such constraints into the optimization problem described above. This would enable the selection of noise parameters in a manner that considers privacy against both estimation and identification using public information.

3.1   Model Complexity

Complex noise models enable superior tradeoffs between privacy and distortion. For example, in a noise model that has one parameter per record, records that are similar to many other records can be perturbed with little noise, while records that are outliers can be distorted more heavily. Naturally, the optimization problem described in (10) becomes more computationally expensive to solve as the noise models increase in complexity. The ability to fit both simple and complex noise models is one of the strengths of our approach, since it enables tradeoffs between model complexity and computational cost.

Figure 2 presents solutions of the problem shown in (10) for noise models of varying complexity. The optimization in this section (and throughout this paper) is performed using the Matlab Optimization Toolbox implementation of sequential quadratic programming. In this example we consider a synthetic database consisting of two hundred records, where each record has three real-valued attributes. Each attribute is generated by sampling from a Gaussian. The Gaussian distribution used to generate each attribute differs in mean and variance. We consider the privacy and distortion tradeoffs achievable by noise models of varying complexity. Figure 2 compares three models of noise perturbation. Each model consists of adding zero mean Gaussian noise, but we vary the variances by record and attribute. In the first model, a single variance characterizes the Gaussian used to perturb all values in the database. This model has one variable. In the second model, we use a different variance for each attribute. This model has three variables. Finally, in our most complex model, one variance models the perturbation for each attribute in each record. This model

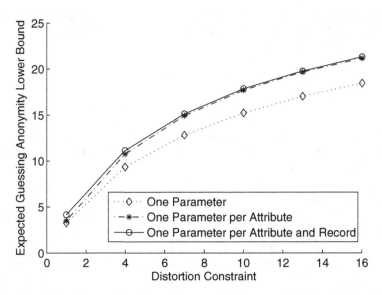

Fig. 2. Distortion for models of varying complexity, as a function of the distortion constraint (constraint that the expected distortion must be less than C)

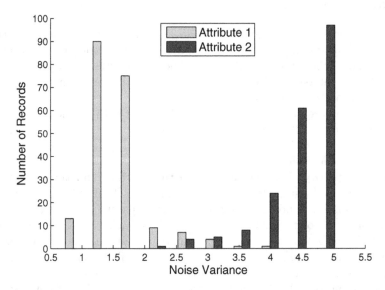

Fig. 3. Histograms for two attributes over the noise parameters assigned to each record in the synthetic data set

has 200 * 3=600 variables. Figure 2 shows a plot of the maximum lower bound on the expected guessing anonymity for various distortion constraints. We used mean squared error as our measure of distortion. As expected, the simplest noise model that relies on a single variable has the lowest privacy for each value of the distortion constraint. Perhaps counter-intuitively, the 3 parameter per-attribute noise model achieves nearly the same level of privacy as the most complex, 600 variable model. This can be explained by the fact that we generated the data by sampling different Gaussian distributions for different attributes, not for different records (which is a common model for real-world data sets). This indicates that the choice of noise model is an important step in achieving privacy through noise perturbation using our approach. Careful choice of the model can yield good results with a significant reduction in computational costs.

Our approach enables the principled use of complex noise-perturbation models to achieve privacy. Such models can achieve higher privacy for the same level of distortion as demonstrated above. This superior performance is a result of the fact that the noise level can be varied depending on whether a record is an outlier or in a dense region of the data. For example, Figure 3 plots a histogram over the variances assigned to different records for two different attributes by the most complex model discussed in the previous paragraph. As expected, the model barely perturbs some records, while heavily perturbing others. Further, the distribution of variances differs sharply by attribute.

3.2   Anonymity Distribution

The theory behind guessing anonymization relies on the guessing inequality, which gives a lower bound on the expected number of guesses necessary for identification. While the lower bound is conceptually important and analytically useful, the expectation does not fully describe the distribution of guessing anonymities across records. In this section, we discuss the empirical distribution of guessing anonymities and the effect of the distortion constraint on number of records with low guessing anonymity.

Figure 4 shows a histogram over the empirical guessing anonymities of an anonymized synthetic dataset consisting of four hundred records and three attributes. Each record was generated by sampling three identical Gaussian distributions. The noise model used was additive zero mean Gaussian noise, with one variance per attribute (i.e. three parameters in total). Solving the optimization problem described in (10) using a distortion constraint of $C = 30$, we obtained parameters used to anonymize the data set. As can be seen from the histogram, there is a wide distribution of guessing anonymity values, ranging from four hundred to one. The number of records that had a guessing anonymity of one was four, or 1% of the data set.

Naturally, one would expect the fraction of records with a small guessing anonymity to decrease as we increased the distortion constraint (i.e. as we allowed more noise). Figure 5 demonstrates this for the four hundred record data set described in the previous paragraph. The figure shows the fraction of records whose empirical guessing anonymity was less than or equal to a constant (one, two, and four for the respective curves) averaged over 200 experiments. The error bars show an interval of two standard deviations. Figure 5 shows that there are very few records with a low empirical guessing anonymity (≤ 1, 2, or 4), and

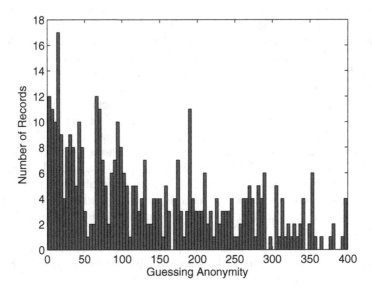

Fig. 4. Histogram over empirical guessing anonymity values for synthetic data set

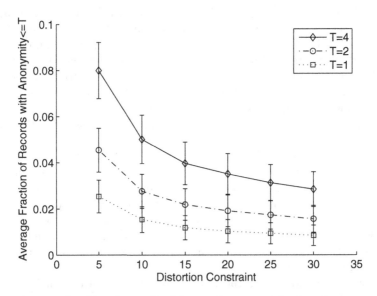

Fig. 5. Average fraction of records with guessing anonymity less than or equal to the given threshold

that this number, as expected, decreases as the distortion constraint increases. To protect records with a low guessing anonymity one could increase the distortion constraint, and thus allow the model more noise with which to anonymize the data. Alternatively, since difficult to anonymize points are often outliers, one could simply remove or alter these points during a second pass after the initial anonymization process. Such a process for eliminating outliers is used in other anonymization methods, e.g., [18].

4   Evaluation

In this section, we evaluate the performance of guessing anonymity-based anonymization using a real data set for two common data mining tasks: classification and regression. We also compare the performance of our approach to [14]. In order to present comparable and replicable results on a publicly available data set, we used the census (adult) data in the UCI Dataset [19]. The UCI adult dataset consists of 14 fields, plus the class to predict. Table 4 contains the fields and sample records from the dataset. Our experiments in Section 4.1 use the full UCI adult dataset (32, 561 records) with thirteen attributes (both real and categorical). We removed the attribute 'fnlwgt' since it only contains information about the data collection process.

Table 4. Sample records from the UCI adult data set. In our experiments, we used all fields of the data, excluding the field 'fnlwgt' which only contains information about the collection process. In classification, we predict income; in regression, we predict age.

age	workclass	education	occupation	race	sex	...	native-country	income
31	Private	Masters	Prof-specialty	White	Female	...	United-States	> 50K
18	Private	HS-grad,	Other-service	White	Female	...	Honduras	<= 50K

4.1   Classification and Regression

We study the effect of data anonymization using our approach on classification and regression performance. To anonymize the Adult data set using our approach we used a combination of discrete and continuous noise models. Our noise model corrupted each real attribute with zero mean Gaussian noise with a different variance for each real attribute. For the discrete attributes, we assumed a simple noise model. For discrete attribute k, the noise model outputs the true attribute value with probability $1 - p_k$, and an incorrect value with probability p_k distributed uniformly over the other discrete values. To optimize this model, we used the Matlab Optimization Toolbox implementation of sequential quadratic programming. Since the noise model contains both real and categorical variables, our distortion metric is a weighted combination of (8) and (9). We weighted the discrete distortion more heavily to provide parity between the scale of the distortion values of the mean-squared error metric and the hamming metric. Figure 6 shows the increase in the expected guessing anonymity lower bound as we increase the distortion constraint.

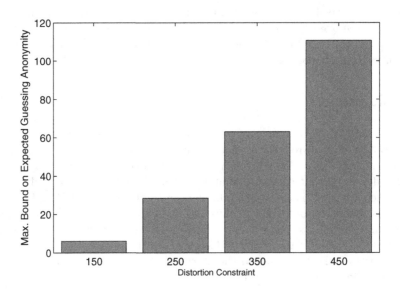

Fig. 6. Maximum guessing anonymity lower bound of the UCI Adult data set as a function of the distortion constraint

For the classification task, we predict the income level for each record. We use J48, a decision tree implementation in Weka [20], with 10-fold cross validation, to run the classification experiments. Figure 7 shows that as distortion and privacy increase, the classification performance decreases. The dashed line represents the classification performance using the un-anonymized data.

For regression, we predict the age attribute using the other attributes including income level. We use linear regression as implemented in Weka, with 10-fold cross validation. Figure 8 shows that as the distortion and privacy levels increase, the correlation coefficient decreases monotonically from the correlation coefficient found using the un-anonymized data. In our experiments, it appears that data distortion is a reasonable proxy for the utility of the data for tasks such as regression and classification.

4.2   Comparison with Probabilistic K-Anonymity

We compare the performance of our approach to probabilistic k-anonymity [14], the most closely related work to the approach presented in this paper. Since [14] only applies to real attributes, we conduct our comparison using the real attributes of the UCI adult data set (excluding the attribute 'fnlwgt'). In our comparison, we assume that the noise perturbation model is a zero mean Gaussian perturbation, with one variance parameter per record.

In order to compare the two approaches, we begin by setting a distortion constraint, and solving the optimization problem described in (10) to find noise parameters that maximize privacy under this constraint. We use these

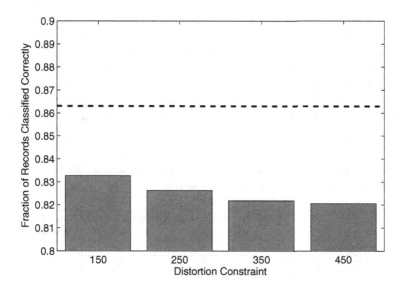

Fig. 7. Empirical percentage of correct classifications when classifying income in the UCI Adult data set as a function of the distortion constraint. The dashed line represents the classifier performance over the original, un-anonymized data set.

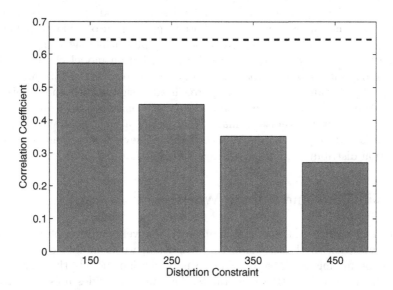

Fig. 8. Empirical correlation coefficient values when predicting age in the UCI Adult data set as a function of the distortion constraint. The dashed line represents the correlation coefficient over the original, un-anonymized data set.

parameters to anonymize the real attributes of five hundred randomly selected records from the adult data set. We empirically determine the anonymity value (as defined in [14]) for each record in the the anonymized data set. The approach described in [14] takes these anonymity values as inputs, and determines the noise parameter required to achieve the desired anonymity value for each record. Given these noise parameters, we anonymize the original data set to obtain a second anonymized set of records. We compare the two anonymized data sets, obtained using the two approaches, by performing a linking attack using the true quasi-identifiers.

Table 5. Comparison between guessing anonymity-based anonymization and the probabilistic k-anonymity-based method presented in [14]

	Number of Records Unprotected	Average Number of Guesses	Average Variance
guessing anonymity	57	68.9	100
prob. k-anonymity [14]	317	19.8	4655

The results of this comparison are shown in Table 5. As can be seen in this table, the guessing anonymity-based approach achieves a higher level of privacy using significantly less noise. Much fewer records were completely unprotected in the data set anonymized using guessing anonymity. A record is completely unprotected if the true record is the first guess in a linking attack on a sanitized record. The guessing anonymity-based approach also achieved a higher average guessing anonymity over all records. Despite achieving a higher level of privacy, the mean variance of the noise perturbation model chosen by the guessing anonymity-based approach is significantly lower. Our results show that the probabilistic k-anonymity-based approach adds significantly more noise to the data. When looking at the results closely, we observed that probabilistic k-anonymity added little noise to many records, but a high amount of noise to some records. Our approach, on the other hand, achieves a more balanced noise perturbation distribution, and on average adds less noise.

5   Conclusions and Future Work

This paper introduces the guessing anonymity, a novel definition of privacy for noise perturbation methods. Using this definition, we formulate optimization problems for finding parameters of noise perturbation models that maximize privacy under data distortion constraints. This work provides noise perturbation methods with privacy guarantees typically associated with k-anonymity, while retaining advantages of noise perturbation methods such as clear statistical assumptions.

Our optimization formulation naturally accommodates both real and categorical variables, and enables the deployment of complex noise models. We apply our approach to the UCI adult data set, and empirically evaluate the impact of anonymization on regression and classification performance. Further, we compare our work to [14], and demonstrate that our approach can yield higher privacy with less distortion of the data.

We are currently pursuing several directions of research based on the work described in this paper. Results on guessing can be used to quantify the effect of attacker side information, such as knowledge of the parameters of the noise model, on anonymity. This would enable us to provide privacy guarantees as a function of assumptions about attacker knowledge. Another direction for future work explores constraining not only the expected guessing anonymity but also its variance. The feasibility of bounding the guessing anonymity variance can be appreciated by noticing that the guessing inequality upper and lower bounds all moments of the guessing function. This would enable noise perturbation methods to provide guarantees not only on the expected guessing anonymity, but also the variance of guessing anonymities. Finally, as shown in the experiments in this paper, the choice of noise model results in a tradeoff between computational complexity and anonymization quality. Selection of noise models and records to manage this tradeoff is an important issue in the application of this approach.

References

1. Barbaro, M., Zeller, T.: A face is exposed for aol searcher no. 4417749, New York Times (August 9, 2006)
2. Domingo-Ferrer, J.: A survey of inference control methods for privacy-preserving data mining. In: Aggarwal, C.C., Yu, P.S. (eds.) Privacy-Preserving Data Mining: Models and Algorithms. Springer, Heidelberg (2008)
3. Aggarwal, C.C., Yu, P.S.: A general survey of privacy-preserving data mining models and algorithms. In: Aggarwal, C.C., Yu, P.S. (eds.) Privacy-Preserving Data Mining: Models and Algorithms. Springer, Heidelberg (2008)
4. Agrawal, R., Srikant, R.: Privacy-preserving data mining. In: ACM SIGMOD Conference (2000)
5. Agrawal, D., Aggarwal, C.C.: On the design and quantification of privacy preserving data mining. In: ACM PODS Conference (2002)
6. Muralidhar, K., Sarathy, R.: Security of random data perturbation methods. ACM Trans. Database Syst. 24(4) (1999)
7. Samarati, P., Sweeney, L.: Protecting privacy when disclosing information: k-anonymity and its enforcement through generalization and suppression. In: Proceedings of the IEEE Symposium on Research in Security and Privacy (1998)
8. Bayardo, R.J., Agrawal, R.: Data privacy through optimal k-anonymization. In: IEEE International Conference on Data Engineering, pp. 217–228 (2005)
9. LeFevre, K., DeWitt, D.J., Ramakrishnan, R.: Incognito: efficient full-domain k-anonymity. In: ACM SIGMOD (2005)
10. Aggarwal, C.C.: On randomization, public information, and the curse of dimensionality. In: IEEE International Conference on Data Engineering (2007)
11. Torra, V., Abowd, J., Domingo-Ferrer, J.: Using mahalanobis distance-based record linkage for disclosure risk assessment. In: Domingo-Ferrer, J., Franconi, L. (eds.) PSD 2006. LNCS, vol. 4302, pp. 233–242. Springer, Heidelberg (2006)

12. Domingo-Ferrer, J., Torra, V.: A quantitative comparison of disclosure control methods for microdata. In: Doyle, P., Lane, J., Theeuwes, J., Zayatz, L. (eds.) Confidentiality, disclosure, and data access: Theory and practical applications for statistical agencies, pp. 111–133. Elsevier, Amsterdam (2001)
13. Arikan, E.: An inequality on guessing and its application to sequential decoding. IEEE Transactions on Information Theory 42(1), 99–105 (1996)
14. Aggarwal, C.C.: On unifying privacy and uncertain data models. In: IEEE International Conference on Data Engineering (2008)
15. Renyi, A.: On measures of entropy and information. In: 4th Berkeley Symposium on Mathematical Statistics and Probability (1961)
16. Dalenius, T.: Finding a needle in a haystack - or identifying anonymous census record. J. Official Statistics 2(3), 329–336 (1986)
17. Massey, J.L.: Guessing and entropy. In: IEEE Symposium on Information Theory (1994)
18. Sweeney, L.: Achieving k-anonymity privacy protection using generalization and suppression. International Journal on Uncertainty, Fuzziness and Knowledge-based Systems 10(5), 571–588 (2002)
19. Asuncion, A., Newman, D.: UCI machine learning repository adult dataset (2007)
20. Witten, I.H., Frank, E.: Data mining: Practical machine learning tools and techniques

Strategies for Effective Shilling Attacks against Recommender Systems

Sanjog Ray and Ambuj Mahanti

Indian Institute of Management Calcutta, Management Information Systems Group,
Joka,700104 Kolkata, India
{Fp062004,Am}@iimcal.ac.in

Abstract. One area of research which has recently gained importance is the security of recommender systems. Malicious users may influence the recommender system by inserting biased data into the system. Such attacks may lead to erosion of user trust in the objectivity and accuracy of the system. In this paper, we propose a new approach for creating attack strategies. Our paper explores the importance of target item and filler items in mounting effective shilling attacks. Unlike previous approaches, we propose strategies built specifically for user based and item based collaborative filtering systems. Our attack strategies are based on intelligent selection of filler items. Filler items are selected on the basis of the target item rating distribution. We show through experiments that our strategies are effective against both user based and item based collaborative filtering systems. Our approach is shown to provide substantial improvement in attack effectiveness over existing attack models.

Keywords: Recommender systems, Shilling attacks, Collaborative filtering.

1 Introduction

Recommender systems are technology based systems that provide personalized recommendations to users. In these systems, opinions and actions of other users with similar tastes are used to generate recommendations. However, with increasing popularity of recommender systems in ecommerce sites they have become susceptible to shilling attacks. In shilling attacks, attackers try to influence the system by inserting biased data into the system. Recent researches have started focusing on attack models and attack detection strategies. [4, 6, and 8]

An attack on a recommender system is mounted by injecting a set of biased attack profiles into the system. Each attack profile contains biased ratings data and a target item. Profiles are injected into the system by fictitious user identities created by the attacker. Every attack can be classified as a push attack or a nuke attack. In a push attack, the objective of the attacker is to increase the likelihood of the target item being recommended to a large section of the users in the system. While in a nuke attack, the objective is to prevent the target item from being recommended.

An attack is also classified as a *high-knowledge* attack or *low-knowledge* attack [8]. A *high-knowledge* attack requires more detail knowledge of the rating distributions of

F. Bonchi et al. (Eds.): PinKDD 2008, LNCS 5456, pp. 111–125, 2009.

each item present in the system. While in a *low-knowledge* attack, to launch an attack, dependence on the recommender system for information on the items is minimal. The approach of constructing the attack profile, based on knowledge about the items, products, and users of recommender systems is known as an attack model. The primary objective of an attacker is to build attack models that provide the most impact with minimal knowledge. The other concern of importance to an attacker is to create models which are hard to detect by attack detection algorithms.

The general form of a push attack profile is shown in figure 1. An attack profile consists of a set of m ratings for m items; where m is the total number of items present in the system. This attack profile of m ratings can be divided into four sets of items: a target item i_t, a set of selected items I_S, a set of filler items usually randomly chosen I_F, and a set of unrated items I_E. While mounting an attack, the set of selected items remains same for all the attack profiles. For example, if a set of 10 attack profiles are inserted into the system to mount a push attack then all the 10 attack profiles will have the same set of selected items while the set of filler items for each profile will differ as they are randomly selected. So, all attack profiles that are inserted into the system to mount an attack will have the same set of selected items and target item. Attack models are defined by the rules by which the four set of items are identified and the way ratings are assigned to the items present in the sets. For some attacks set of selected items may be empty.

Selected Items (I_S)	Filler items (I_F)	Unrated items (I_E)	Target item (i_t)

Fig. 1. A general form of attack profile

Most past researches have mainly focused on the set of selected items while creating new attack models. Earlier attack models have been proposed without a discussion of their success chances in a real world scenario. In this paper, we discuss the importance of target item as a critical factor in the constructing of effective shilling attacks. We propose attack strategies based on target item rating distribution. Our attack strategies also examine the importance of fillers items in improving the effectiveness of an attack. Unlike previous approaches, our attack strategies are tailored on the basis of the recommender system algorithm attacked. Our paper provides different attack models for mounting effective attacks against user based collaborative filtering systems and item based collaborative filtering systems respectively. This work is a novel approach that is focused on intelligent use of filler items based on target item rating distribution. Through experimental evaluation we show that our proposed strategies can result in more effective attacks against both user based and item based recommender systems.

This paper is organized as follows. In section 2 we provide a brief summary of various attack models and their filler strategies. A discussion on target item selection and classification approach is presented in section 3. In section 4 we describe user based and item based collaborative filtering algorithm. Evaluation metric used is also described in section 4. In section 5 and section 6 we provide details of our proposed filler items strategies for user based and item based collaborative filtering systems respectively. In section 7 we describe the experimental evaluation process and report the results obtained. Finally, we conclude the paper in section 8.

2 Types of Attack

2.1 Example of Attack against User Based Collaborative Filtering

Before describing different existing attack models, we first illustrate an example that shows the consequence of a shilling attack against a recommender system implementing user based collaborative filtering. In user based recommender systems, recommendations are generated based on the ratings given to the target item by users similar to the target user. Consider a recommendation system for movies. Let Alice be a user in this system, whose rating for item 5 (an item not yet seen by Alice) is to be predicted. Table 1 displays Alice profile with 5 genuine users. If the system predicts the rating given to a target item by Alice on the basis of the rating given to the target item by the user most similar to Alice, then from table 1 we can conclude that the predicted rating for item 5 in case of user Alice will be 2. Because among all users who have rated the target item, user 5 has the highest similarity with Alice and he has rated item 5 as 2. It can also be observed that even though user 4 has the highest similarity with Alice its ratings cannot be used while generating predictions for item 5 as user 4 has not rated item 5.

Now let us consider that an attacker Eve inserts two attack profiles into the system with the purpose of promoting item 5. Table 2 shows the new correlations of Alice with the 5 genuine users and the 2 attack profiles after the attack. We can observe that now the predicted rating for item 5 for Alice will be 5 because attack 1 profile has the highest correlation with Alice after the attack. This predicted rating of 5 is exactly the opposite of the rating earlier predicted for Alice towards item 5. This example clearly shows how a successful shilling attack can result in undeserving items being promoted to genuine users. This can result in customer dissatisfaction and result in lowering of customer trust on the ability of the recommender system to make better predictions.

Similarly, in case of item based recommendation systems, shilling attacks can result in target item prediction been manipulated to help the attacker achieve his goal. Item based systems are very similar to the one explained in this example. In item based systems, similarities are calculated between items instead of users. As it is difficult to manipulate the similarities of two items compared to that of two users, item based recommender systems are found to be more robust that user based systems [8]. A more detail explanation of the two systems can be found in section 4.

Table 1.

	Item 1	Item 2	Item 3	Item 4	Item 5	Correlation With Alice
Alice	4	4	1	3	?	
User 1	2	3		2	1	0.5
User 2	4	1	2		2	0.1889
User 3		3	4	4	4	-0.7559
User 4	4		1	3		1
User 5	3	4	3	4	2	0.4082

Table 2.

	Item 1	Item 2	Item 3	Item 4	Item 5	Correlation With Alice
Alice	4	4	1	3	?	
User 1	2	3		2	1	0.5
User 2	4	1	2		2	0.1889
User 3		3	4	4	4	-0.7559
User 4	4		1	3		1
User 5	3	4	3	4	2	0.4082
Attack 1	4	4	3	4	5	0.9428
Attack 2	4	3	2	4	5	0.7385

2.2 Attack Models

Various attack models have been proposed in previous researches on shilling of recommender systems. We discuss below, some of the popular attack models on which much research is focused on. The attack models described in this paper are explained in context of a push attack. A comprehensive study of different attack models can be found in [8].

Random attack: Random attack is the most simplest of all attack models. In this model, filler items are selected randomly from the set of all items. Filler items are then assigned ratings selected from a set of random values chosen from a distribution centered on the system mean. System mean is the mean for all user ratings across all items. This is *a low – knowledge* attack as minimal knowledge is required to obtain system mean value. It has been found that this model is not very effective [8].

Average attack: In an average attack model, set of selected items is empty. Filler items are selected randomly, and each filler item is assigned its mean rating. Mean rating here corresponds to the average rating for the item across all users in the database who have rated it. In terms of attack effectiveness, it is one of the most powerful attack models. Average attack is a *high-knowledge* attack as mean rating of each filler item is required to mount an attack. However, in [1] it has been shown that this attack can be effective with limited knowledge i.e. a small set of filler items can perform an effective attack.

Bandwagon attack: In this model, the set of selected items contains few of those items that have high popularity among users. Thus, attack profiles created will have higher chances of being similar to a large number of users. Selected items and the target item are assigned maximum rating value. As in random attack, filler items are randomly selected and assigned mean rating of items across the whole system. Therefore, bandwagon attack can be seen as an extension of the random attack. Bandwagon attack is a *low–knowledge* attack as popular items data can be obtained from publicly available information sources.

Segmented attack: This attack is modeled to push the target item to those users who are most likely to be influenced by the recommendation. A segment is defined as a group of users having affinity for items of similar features. Group of users who have rated highly most of the popular horror movies is an example of a segment of users interested in horror movies. An attacker with intent to promote a horror movie will try to get his target item recommended to this segment of users as the likelihood of influencing them is higher. In this model, the set of selected items contains few of those items that have high popularity among users of the targeted segment. Selected items and the target item are assigned maximum rating value. Filler items are identified randomly and given the lowest possible rating. It has been shown that segmented attack is the most effective model against in-segment users. It is a *low-knowledge* attack as selection of highly rated movie with similar features can be achieved from public information sources.

3 Target Item Selection and Classification

3.1 Target Item Selection

Attack models described earlier have never considered the importance of selecting the right target item for shilling. The reason is that attack models were designed to take into consideration only the user ratings data. We propose that every item present in the database should not be considered as the right candidate for shilling. By right candidate we mean those items that have higher chances of being selected by a user when presented as a recommendation. The factors which need to be considered for selecting the right target item can be discovered by studying the way a user selects a movie for viewing or renting. One factor which we believe plays a important role in influencing a user buying decision of a item like movie or book are the ratings and reviews given to the particular item by other customers. In [5], a study of user behavior in Amazon online store shows that customers reviews and ratings plays an important role in a prospective buyer buying decision. We also believe that ratings and reviews expressed on the target item by experts or other users in popular sites other than the site where the target user conducts the transaction also play a vital role in the buying decision. For example a person buying a movie from Amazon.com may also be influenced by the ratings expressed for the movie in popular website like IMDB.com.

While selecting a target item for shilling, the reviews and ratings given to the item by customers and experts should be taken into consideration, because a target item which has been given low ratings and bad reviews by most customers and expert reviewers will have low probability of being selected by a target user. A shilling attack may push an item disliked by both customers and expert reviewers into a target user recommendation list but the chances of it been selected by the target user are low.

Our proposed approach is to select those movies as target item that are debatable or those movies which are liked by all. Debatable item implies an item which has mixed ratings from both customers and reviewers i.e. items which have no strong consensus about its likeability. Target items which are liked by all are also good candidates for shilling attacks as the chance of the target user buying it is higher. Items with low ratings and that are disliked by all should be avoided as the probability of it influencing a target user is very low. Our recommendation of selecting a debatable item or an item already rated highly by other users as the ideal target item is based on the assumption that a user is more likely to buy an item higher up in the recommendation list. For example, if an attack is mounted on a popular item, it may move the target item to position 2 from position 8 in a top-10 ordered recommendation list for a user. Our assumption is that the user has a higher probability of selecting the item at position 2 in the recommendation list as compared to the item in position 8.

3.2 Target Item Classification

Our attack strategies are dependent on target item rating distribution. Target item is categorized into T_L or T_H category on the basis of the target item rating distribution.

T_L: Target item with majority of their ratings at lower end of the rating scale fall into this category. In our experiment we grouped items with 60 % or more of their ratings as 1 or 2 in this category.
T_H: Target item with majority of their ratings at higher end of the rating scale fall into this category. In our experiment we grouped items with 60 % or more of their ratings as 4 or 5 in this category.

Once category of the target item and the recommendation system algorithm to be attacked is known, appropriate strategies based on filler items are then used to construct attack profiles.

4 Recommendation Algorithm and Evaluation Metric

In this paper we have evaluated our filler item attack strategies against both user based collaborative filtering algorithm and item based collaborative filtering algorithm. In collaborative filtering (CF), a user is recommended items that people with similar tastes and preferences liked in the past. This technique mainly relies on explicit ratings given by the user and is the most successful and widely used technique [3]. In this section we describe the collaborative filtering algorithms, the evaluation metric used, and the notion of prediction shift.

4.1 User Based Collaborative Filtering

In user based collaborative filtering [2], firstly, neighborhood of k similar users is found for the target user. Then for generating prediction for an item not yet seen by the target user, weighted average of the ratings given by the k similar neighbors towards the predicted item is used.

To calculate similarity among users we use Pearson-r correlation coefficient. Let the set of items rated by both users u and v be denoted by I, then similarity coefficient ($Sim_{u,v}$) between them is calculated as

$$Sim_{u,v} = \frac{\sum_{i \in I}(r_{u,i} - \bar{r_u})(r_{v,i} - \bar{r_v})}{\sqrt{\sum_{i \in I}(r_{u,i} - \bar{r_u})^2}\sqrt{\sum_{i \in I}(r_{v,i} - \bar{r_v})^2}} \qquad (1)$$

Here $r_{u,i}$ denotes the rating of user u for item i, and $\bar{r_u}$ is the average rating given by user u calculated over all items rated by u.

Similarly, $r_{v,i}$ denotes the rating of user v for item i, and $\bar{r_v}$ is the average rating given by user v calculated over all items rated by v.

To compute the prediction for an item i for target user u, we use the following formula.

$$P_{u,i} = \bar{r}_u + \frac{\sum_{v \in V} Sim_{u,v}(r_{v,i} - \bar{r}_v)}{\sum_{v \in V} |Sim_{u,v}|} \qquad (2)$$

Where V represents the set of k similar users. While calculating prediction only those users in set V who have rated item i are considered.

4.2   Item Based Collaborative Filtering

In item based collaborative filtering [10], similarities between the various items are computed. From the set of items rated by the target user, k items most similar to the target item are selected. For computing the prediction for the target item, weighted average is taken of the target user's ratings on the k similar items earlier selected. Weightage used is the similarity coefficient value between the target item and target user items.

To compute item-item similarity we used adjusted cosine similarity. Let the set of users who rated both items i and j be denoted by U, then similarity coefficient ($Sim_{i,j}$) between them is calculated as

$$Sim_{i,j} = \frac{\sum_{u \in U}(r_{u,i} - \bar{r}_u)(r_{u,j} - \bar{r}_u)}{\sqrt{\sum_{u \in U}(r_{u,i} - \bar{r}_u)^2}\sqrt{\sum_{u \in U}(r_{u,j} - \bar{r}_u)^2}} \qquad (3)$$

Here $r_{u,i}$ denotes the rating of user u for item i, and $\bar{r}_u$ is the average rating given by user u calculated over all items rated by u.

Similarly, $r_{u,j}$ denotes the rating of user u for item j.

To compute the predicted rating for a target item i for target user u, we use the following formula.

$$P_{u,i} = \frac{\sum_{j \in I} Sim_{i,j} * r_{u,j}}{\sum_{j \in I} |Sim_{i,j}|} \qquad (4)$$

In equation 4, I represent the set of k most similar items to target item i that have already been rated by the target user u. As earlier mentioned, $r_{u,j}$ denotes the rating of user u for item j.

4.3   Prediction Shift

For the purpose of measuring the effectiveness of the attack we use the widely used metric prediction shift. Prediction shift of a target item is the difference of average predicted rate of the target item, after and before the attack, for all target users. Average Prediction shift of an attack is the average change in prediction for all target items. We use the same formula as in [8], which is defined as follows.

Let U and I be the sets of target users and target items. Let $\Delta_{u,i}$ denote the prediction shift for user u on item i. $\Delta_{u,i}$ can be measured as $\Delta_{u,i} = p'_{u,i} - p_{u,i}$, where $p'_{u,i}$ is the prediction value after the attack and $p_{u,i}$ before the attack. The average prediction shift for an item i over all users can be computed as

$$\Delta_i = \frac{\sum_{u \in U} \Delta_{u,i}}{|U|} \tag{5}$$

The prediction shift for an attack model is the average prediction shift for all items tested. It can be computed as

$$\Delta = \frac{\sum_{i \in I} \Delta_i}{|I|} \tag{6}$$

5 Filler Item Strategies for User Based Collaborative Filtering

Known attack models like average attack, bandwagon attack, and segmented attack are focused at creating attack profiles which have greater chance of having high similarity with as many users as possible. Because an attack profile with high similarity with a genuine user increases the chance of it being selected in top k similar neighbors of the user, thereby influencing the rating of the target item. Bandwagon attack and segmented attack have tried to achieve this objective by selecting popular items as part of their attack profiles. An attack profile consisting of the popular items will have similarity with higher number of users, which should finally result in an effective attack. However, it has been found that average attack is more effective compared to bandwagon and segmented attack [8]. One possible reason for this unexpected result could be the presence of other factors which also affect the effectiveness of an attack.

Our proposed approach for user based CF systems considers rating distribution of the target item as a critical factor that can affect the effectiveness of an attack. Unlike previous attack models which focus on creating attack profiles that are similar to as many users as possible from the set of all users; the objective of our approach is to create attack profiles that increase similarity with as many of those users who have rated the target item. Our approach proposes two strategies that are based on the rating distribution of the target item. Both strategies improve similarity by assigning appropriate values to filler items. As we show below, our proposed approach performs substantially better than average attack model. We define below the two strategies.

5.1 Strategy UL

This strategy is followed when the target item falls in T_L category. As majority of the users have rated the target item at the lower end of the rating scale, to improve effectiveness of the attack, we need to create profiles that are similar to the users who have rated the target item with a lower value. So, to improve similarity, a randomly selected filler item is assigned the average rating given to the filler item by those users who have rated the target item at the lower scale.

Let U_A be the set of all users who have rated the randomly selected filler item I_F. Let U_L be the set of all users who have rated the target item at a lower scale of rating and have also rated the randomly selected filler item. So, in this strategy, filler item I_F is assigned the average rating given to it by the set of users U_L. This approach differs from average attack in the way filler items are assigned values, as in an average attack item I_F was assigned the average rating given to it by the set of users U_A.

5.2 Strategy UH

This strategy is followed when the target item falls in T_H category. As a large majority of the users have rated the target item at the higher end of the rating scale, to improve effectiveness of the attack, we need to create profiles that are similar to the users who have rated the target item with a higher value. So, to improve similarity, a randomly selected filler item is assigned the average rating given to the filler item by those users who have rated the target item at the higher scale.

Let U_A be the set of all users who have rated the randomly selected filler item I_F. Let U_H be the set of all users who have rated the target item at a higher scale of rating and have also rated the randomly selected filler item. So, in this strategy, filler item I_F is assigned the average rating given to it by the set of users U_H. This approach differs from average attack in the way filler items are assigned values, as in average attack item I_F was assigned the average rating given to it by the set of users U_A.

6 Filler Item Strategies for Item Based Collaborative Filtering

It has also been found that attack models are not as effective against item based CF systems as they are against user based CF systems [8]. One explanation of this may be because most attack models are designed to improve similarity among users than items.

Our proposed approach for item based CF system is designed to improve similarity among items and considers rating distribution of the target item as a critical factor that can affect the effectiveness of an attack. Our approach proposes two strategies that are based on the rating distribution of the target item. Both strategies improve similarity by selecting appropriate filler items. While in filler item strategies against user based CF systems we had modified an existing attack model i.e. average attack model by intelligently assigning values to filler items, the strategies elaborated in this section is a new approach that is focused on selection of appropriate filler items and is specifically built for attack against item based collaborative filtering systems. As we show below, our proposed approach performs substantially better than average attack model. We define below the two strategies.

6.1 Strategy IL

This strategy is followed when the target item falls in T_L category. As majority of the users have rated the target item at the lower end of the rating scale, it is most likely that items similar to the target item will also be rated at the lower end of the scale. To improve effectiveness of the attack, we need to create profiles that increase the similarity of the target item with items which are rated at the higher end of the rating scale. So, to improve similarity, we select filler items from the set of items which are highly rated by those users who have rated target item at a lower scale. In a scale of 1 to 5, higher scale implies rating of 4 or 5, lower scale implies a rating of 1 or 2.

Let U_L be the set of all users who have rated the target item at a lower scale of rating i.e. 1 or 2. Let I_F be the set of items that have been rated 4 or 5 by the set of users U_L. So, in this strategy, filler items are selected from this set of items I_F. To select filler items, frequency count of all items which are rated 5 in set I_F is computed and those with higher frequency count are given preference while selecting filler items. In case number of filler items required for creating a attack profile cannot be fulfilled by all the 5 rated items in the set I_F then a frequency count of all 4 rated items in set I_F is computed. As in earlier case, during selection of filler items, items with higher count are given preference. It is also taken into consideration that filler items selected should be distinct. All items selected as filler items are assigned the maximum rating of the rating scale i.e., 5 in a scale of 1 to 5. In case number of filler items required are more than the number items present in set I_F then remaining filler items are selected randomly from the set of all items as done in an average attack. In an average attack, filler items are randomly selected and rating assigned to a filler item is its average rating.

6.2 Strategy IH

This strategy is followed when the target item falls in T_H category. As majority of the users have rated the target item at the higher end of the rating scale, it is most likely that items similar to the target item will also be rated at the higher end of the scale. In this strategy we try to further increase the association of target item with more highly rated items. So, to improve similarity, we select filler items from the set of items which are highly rated by those users who have rated the target item at a higher scale. This will help further strengthen the similarity of target item with highly rated items.

Let U_H be the set of all users who have rated the target item at a higher scale of rating i.e. 4 or 5. Let I_F be the set of items that have been rated 4 or 5 by the set of users U_H. So, in this strategy, filler items are selected from this set of items I_F. The process of selecting filler items from the set I_F is exactly similar to that of strategy L.

The two strategies Strategy IL and Strategy IH differ in the way members of the set U_L and U_H are selected.

7 Experimental Evaluation and Discussion

We performed the experimental evaluation of our strategies on the publicly available MovieLens data set [9]. This is the most widely used dataset in recommender systems research. MovieLens consists of 100,000 ratings made by 943 users on 1682 movies. Each user in the data set has rated at least 20 movies and each movie has been rated at least once. A timestamp value is associated with each user, movie, and rating combination. The data set also contains information on the demographic detail (age, sex, occupation, and zip code) of each user and basic information (genre and release date) of each movie. The ratings are made in a scale of 1 to 5, where a higher rating implies greater likeness for an item.

We evaluated effectiveness of the proposed strategies on user based and item based collaborative algorithm. For similarity calculation and prediction in user based CF algorithm, equations 1 and 2 stated in section 4 were used. Similarly, equations 3 and 4 stated in section 4 were used for computing similarity and prediction value for item

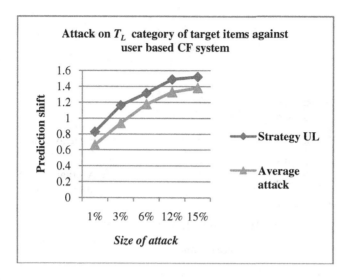

Fig. 2. Attack on T_L category of items against user based collaborative filtering system

based CF algorithm. We used a neighborhood size of $k = 20$ for prediction calculation. Case amplification value of 10 was used while calculating correlation and only positive correlations values were considered for computing predictions.

To conduct our evaluation, we selected a sample 20 items. Out of the 20 items, 10 items belonged to T_L category while remaining 10 items to T_H category. All the 20 items were selected randomly from a larger set of items belonging to each category. We also randomly selected a sample of 50 target users. Target users selected were those who have never rated any of the 20 test items. Each of the target items was attacked individually and the prediction shift was calculated by averaging the prediction shift observed for each user. The final prediction shift for the attack is the average prediction over all items used in the test. Equation 6 was used to calculate the metric.

All experiments were conducted for "*Size of attack*" values 1%, 3%, 6%, 12%, and 15%. "*Size of attack*" represents number of attack profiles added as a percentage of pre-attack profiles. 1% "*Size of attack*" implies 10 attack profiles were added to a system of 1000 genuine users. On the basis of the results reported in [1] that best results are reported when a filler size of 3% is used in an average attack, we used a filler size of 3% for all our tests i.e. 3 % of 1682 items which is approximately 50 filler items. For attacks against user based collaborative filtering systems we used three strategies: Strategy UL, Strategy UH and average attack. Similarly, for attacks against item based collaborative filtering systems we used three strategies: Strategy IL, Strategy IH and average attack. For average attack, filler item strategy used was the same as in an average attack i.e. the mean of the filler item was assigned to it. Category $T_{L,}$ Category $T_{H,}$ Strategy UL, Strategy UH, Strategy IL and Strategy IH were implemented the way explained earlier in section 3, section 5, and section 6 respectively. For attacks against item based CF while selecting filler items from set I_F, only items with minimum frequency count of 10 were considered.

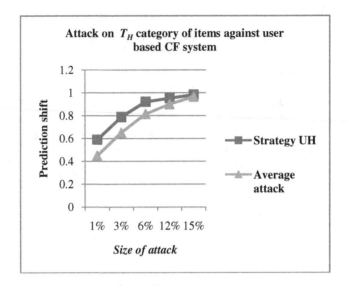

Fig. 3. Attack on T_H category of items against user based collaborative filtering system

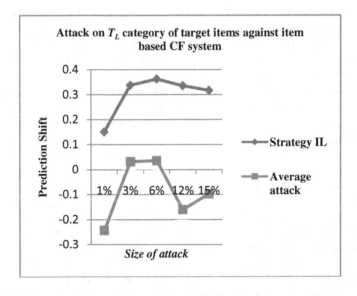

Fig. 4. Attack on T_L category of items against item based collaborative filtering system

Figure 2 and figure 3 show the results of our experiment against systems using user based collaborative filtering for recommendations. Figure 2 shows the prediction shift values of attacks Strategy UL and average attack for items belonging to T_L category. From the graph it's obvious that for items in T_L category, Strategy UL outperforms average attack model for all values of attack size. Similarly, figure 3 shows the prediction shift values for the attack strategies Strategy UH and average attack for items belonging to T_H category. From the graph it can be concluded that for items belonging to T_H category, Strategy UH performs much better than average attack over lower values of attack size. At attack size of 12 % and 15% both attack have similar effectiveness.

Figure 4 and figure 5 show the results of our experiment against systems using item based collaborative filtering for recommendations. Figure 4 shows the prediction shift values of Strategy IL and average attack for items belonging to T_L category. Similarly, figure 5 shows the prediction shift values for the Strategy IH and average attack for items belonging to T_H category. From figure 4 and 5 it can be concluded that both Strategy IL and Strategy IH perform substantially better than average attack over all values of attack size. It can also be observed that our attack strategies are more effective against item based systems than user based systems.

Experimental results clearly show that our approach of selecting a strategy based on target item rating distribution outperforms the best available attack model i.e. average model. One drawback of our attack strategies is its high knowledge cost. However, automated software agents can help diminish the cost. One approach that can be used to decrease the cost is to use a subset of users while selecting filler items. For example, in attacks against item based systems, while implementing Strategy IH instead of selecting all users who have rated target item as 4 or 5 as members of the set U_H. , we only select 20 users. Selection of items for set I_F will then be performed using the data of the 20 users in set U_H. Similarly, in case of attacks against user based systems, while implementing Strategy UH instead of assigning a filler item I_F the average rating given to it by the set of users U_H. , we assign I_F the average rating given to it by a subset of 5 randomly selected users from U_H. In future work we plan to experimentally verify the effectiveness of these cost reduction approaches.

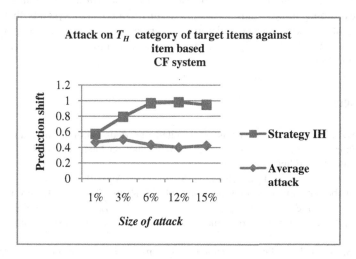

Fig. 5. Attack on T_H category of items against item based collaborative filtering system

In section 3, we have proposed selecting a debatable item or popular item as a target item. We considered an item debatable when it has mixed ratings and comments among users in the ecommerce site and also among reviewers and customers of other popular sites in the target item domain. As our experimental dataset does not have the data that can categorize an item as debatable or popular by analyzing the ratings, comments and reviews of website users and experts, we are unable to experimentally verify the validness of our proposed approach for selecting a target item. In future we plan to define a metric that can be used to measure the suitability of a target item for shilling.

8 Conclusion

This paper examines different strategies that result in effective shilling attacks against both user based and item based collaborative filtering systems. Our novel approach provides a new perspective of mounting attack models. We show the importance of target item and filler item in construction of successful attack strategies. Through experiments we show that our novel approach of intelligent selection of filler items based on target item rating distribution results in substantial improvement over the baseline average attack. We also advocate an approach towards intelligent selection of target item. The motivation behind the approach is to select those target items which have higher probability of selection when presented as a recommendation. One assumption on which our target item selection approach is based on is that users have a higher probability of buying items that are higher up in the recommendation list. In future, we plan to conduct a study to verify this assumption. We also plan to examine the filler items strategies for other attack models, and create algorithms to improve robustness and stability of recommender systems against shilling attacks. While most of the existing work on security of recommender systems are primarily based on developing algorithms that help in identification of attack profiles, we believe the solution of developing more secure systems lies in recommender systems that generate recommendation taking into consideration other parameters like trust, expert reviews etc. In future, we plan to study these factors and their impact in the design of secure recommendation systems.

References

1. Burke, R., Mobasher, B., Bhaumik, R.: Limited Knowledge Shilling Attacks in Collaborative Filtering Systems. In: Proceedings of Workshop on Intelligent Techniques for Web Personalization, Edinburgh (2005)
2. Herlocker, J., Konstan, J., Borchers, A., Riedl, J.: An Algorithm Framework for Performing Collaborative Filtering. In: Proceedings of SIGIR, pp. 77–87. ACM, New York (1999)
3. Konstan, J., Miller, B., Maltz, D., Herlocker, J., Gordon, L., Riedl, J.: GroupLens: Applying Collaborative Filtering to Usenet News. Communications of the ACM 40(3), 77–87 (1997)
4. Lam, S., Riedl, J.: Shilling Recommender Systems for Fun and Profit. In: Proceedings of the 13th International WWW Conference, New York (2004)

5. Leino, J., Raiha, K.: Case Amazon: Ratings and Reviews as Part of Recommendations. In: Proceedings of the 2007 ACM Conference on Recommender Systems, Minneapolis, pp. 137–140 (2007)
6. Mehta, B., Hofmann, T., Nejdl, W.: Robust Collaborative Filtering. In: Proceedings of the 2007 ACM Conference on Recommender Systems, Minneapolis, pp. 49–56 (2007)
7. Mobasher, B., Burke, R., Bhaumik, R., Williams, C.: Effective Attack Models for Shilling Item-Based Collaborative Filtering Systems. In: Proceedings of the 2005 WebKDD Workshop, Chicago (2005)
8. Mobasher, B., Burke, R., Bhaumik, R., Williams, C.: Towards Trustworthy Recommender Systems: An Analysis of Attack Models and Algorithm Robustness. ACM Transactions on Internet Technology 7, 23, 1–38 (2007)
9. MovieLens data set, http://www.grouplens.org
10. Sarwar, B., Karypis, G., Konstan, J., Riedl, J.: Item-based Collaborative Filtering of Recommendation Algorithms. In: Proceedings of the 10th International WWW Conference, Hong Kong (2001)

Author Index